D0467985

SIMON FLEET

CLOCKS

OCTOPUS BOOKS

Listed below is the majority of good public collections of clocks throughout the world. A list of the many private collections is not given, but these can all be ascertained by joining one of the following two societies: The National Association of Watch and Clock Collections Inc., c/o Earl T. Strickler, 335 North Third Street, Columbia, Penna, U.S.A.; or The Antiquarian Horological Society, 35 Northampton Square, Finsbury, London ECI. Both are international in their scope

Ashmolean Museum, Oxford
Boston Museum of Fine Arts; British Museum, London
Carnegie Museum, Pittsburg; Château des Monts, Le Locle
Chigi Collection, Siena; Clock Museum, Leyden
Clockmakers' Company Museum, London Guildhall
Clock Manor, Denver
Conservatoire des Arts et Métiers, Paris
Colonial Museum, Williamsburg
Danske Uhrmuseum, Copenhagen; Deutsches Museum, Munich
Essex Institute, Salem, Mass
Fitzwilliam Museum, Cambridge
Historical Clockwork Museum, Furtwangen
Henry Ford Museum, Dearborn, Mich
James Arthur Collection, New York University
Kunsthistorische Museum, Vienna
Landesmuseum, Cassel; L. G. A. Zentrum, Stuttgart
Louvre, Paris
Moyse's Hall Museum, Bury St Edmunds
Metropolitan Museum, New York
Mathematische Physikalische Salon, Dresden
Musée d'horologie, La Chaux-de-fonds
Musée des Beaux-Arts, Besançon
Musée d'horologie, Geneva
Musée des Augustins, Toulouse
Mody Collection, Tokyo
National Maritime Museum, Greenwich
Nordiska Museet, Stockholm
National Museum, Washington, D C
Nationalmuseum, Würtzburg
Nationalmuseum, Copenhagen
Royal Scottish Museum, Edinburgh
Rosenborg Slot, Copenhagen
Science Museum, London
Smithsonian Institute, Washington
Victoria and Albert Museum, London
Wallace Collection, London

Preceding page
Part of an engraving of the Wells Cathedral clock, see page 84

Acknowledgments

Colour photographs by John Hedgecoe, apart from those listed below
Her Majesty the Queen, figure 67
The Trustees of the British Museum, figures 11, 20, 29, 32, 33, 46, 47, 48, 49, 54, 62, 82, 84, 89, 91, 93, 96, 98, 100, 102, 106, 114, 115, 124, 125, 127
The Victoria and Albert Museum, London, figures 25, 28, 30, 35, 36, 37, 38, 39, 50, 51, 53, 55, 57, 58, 60, 65, 68, 69, 70, 77, 81, 85, 86, 88, 92, 111, 117
The Science Museum, London, figures 1, 3, 4, 5, 7, 9, 13, 16, 17, 18, 22, 31, 40, 43, 45, 52, 80, 95, 104, 105, 107, 108, 110, 113, 129
The Clockmakers' Company Museum, London, figures 8, 12, 14, 24, 26, 90
The National Maritime Museum, figure 103
The Vatican Museum, figure 74
Conservatoire des Arts et Métiers, Paris, figures 19, 44 (Photo by Giraudon)
The Hermitage Museum, Leningrad, figure 78
The Academy San Fernando, Madrid, figure 41
The Whitefield House Collection, Guilford, USA, figure 132
The Burlington Magazine, figure 61
The Chapter of Salisbury Cathedral, figure 119
The Witt Library, Courtauld Institute of Art, end-papers
Dr Frank Tait, figure 94 (photograph by William Henderson);
Felix Harbord Esq, figure 59; Mr and Mrs A. Bryden-Brown, figure 64
M. Pierre Martini, figure 56; Country Life Ltd, figure 67
Godfrey Bonsack Ltd, figure 123; Blairman Ltd, figure 79; Hausmann et Cie, figure 74; E. Hollander Ltd, figure 71; Sotheby's, figure 133
Christies, figure 97
Imprimeries Paul Dupont, figure 2; Faber and Faber Ltd, figure 101 from Snowman's *Art of Carl Fabergé*; E. and F. N. Spon, figures 63, 66, 72, 75, 76 from Britten's *Old Clocks and Watches* 1911; the *New Yorker*, figure 21; Time Life Inc, figure 135; *The Financial Times*, figure 130 (photograph by Geoffrey Buggins); *The Observer*, figure 99 (photograph by Tessa Grimshaw); The Radio Times Hulton Picture Library; Keystone Press Agency, 116, 120, 122, 128, 131, 134; Central Press Photos Ltd
The author is indebted to Dr F. A. B. Ward, of the Science Museum, London, and to Mr Philip Coole of the British Museum, for their assistance and advice

This edition first published 1972 by
OCTOPUS BOOKS LIMITED
59 Grosvenor Street, London W.1

ISBN 7064 0036 4

PRODUCED BY MANDARIN PUBLISHERS LIMITED AND PRINTED IN HONG KONG

Early Time-keepers

1 A garden sundial of 1718

Creep, shadow, creep; my ageing hours tell.
I cannot stop you, so you may as well.
HILAIRE BELLOC

IF WE DO NOT KNOW EXACTLY how many thousands of years ago someone first noticed that the shadow cast by a tree or even by himself pointed in a different direction at different times of day, we know that time was first calculated by the movement of the shadow, and that a stick in the ground was man's first clock [figure 2]. This primitive arrangement was greatly improved two thousand years ago by that nation founded by Moses' great-grandson Nimrod, the Chaldeans, one of the wiser peoples of ancient times, who first divided the day and night into twelve hours each. They placed the stick in a large scooped-out stone, which made the shadow more precisely readable [figure 3]. Examples have been dug up in places as far

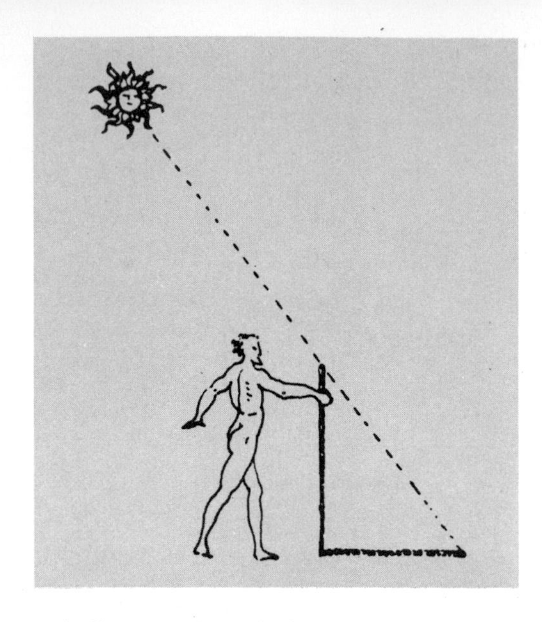

2 A diagram of man's first sundial

3 A plaster cast of a Roman 'hemicycle sundial

apart as Rome and the territory of the Incas in South America.

One of the earliest references to sundials is in the Bible: 'And Isaiah the prophet cried unto the Lord who brought the shadow ten degrees backward by which it had gone down in the dial of Ahaz' (KINGS II, XX, 11). This was in 741 BC. Indian Fakirs and Tibetan priests used special sundial sticks [figure 4]. These had a hole at the top into which a short peg was inserted at a right angle. The stick was then held up by a string at the end and the shadow cast by the peg on the side of the stick showed the time. In ancient Greece time was sometimes calculated by the length of shadow thrown by a column, and measured in terms of a man's steps. Thus a character in Aristophanes' comedy, *The Frogs*, says: 'When the shadow is ten steps long, come to dinner'.

Ruins of monumental sun-clocks stand lonely across the earth. These include the rings of megalithic stones at Stonehenge on Salisbury Plain (*c.* 2000 BC); the vast curved stones at Jaipur in India; the rock pillars in Peru;

the Aztec calendar-clocks [figure 21]; and the famous Egyptian obelisks, some of which cast their shadow round a dial marked in the earth, which is the apotheosis of man's initial stick-in-the-ground-clock. An obelisk sundial of this sort and size was put up some centuries later by the Emperor Augustus in 27 BC on the Campus Martius in Rome, which tells the time there to this day [figure 11]. But the ancient Egyptians invented other shadow-clocks, one being a T-shaped contrivance, a stick fixed at right angles to, and above, another longer stick on which it cast its shadow [figure 7]. The hours were marked by specially placed nails.

In the Middle Ages the sun was exploited for time keeping in yet another way. An aperture was built in the roof of a building, through which a beam of sunlight made a spot on a dial marked out on the floor. There was one like this at Strasbourg Cathedral in the eleventh century, and others are still working at the cathedral in Milan, and at a convent in Catania in Sicily. The highest ever known was made in 1467 at Santa Maria Dei Fiori at Florence. Another form of sundial had a large pin fixed at an angle to a dial; examples can be seen on walls of old houses, churches and cathedrals [figures 6 and 9], and on pedestals in gardens and public parks [figure 1], Boston and Peking each have especially interesting public park sundials. There is a famous one at Chartres cathedral, and an eighteenth-century variety graces the walls of London's Old Bailey; while still to be seen intermittently along the road from Moscow to Leningrad are the sundials which were placed on the milestones for the benefit of travellers by Catherine the Great. At the end of the fifteenth century Margaret of Burgundy had an unusual sundial built in front of the church at Brou. It was without a pointer and the shadow was made by the observer himself. Another dial of this sort has been placed recently outside a school at Basle. However, sundials and sunclocks have one overwhelming fault; they are obviously of little use in cloudy weather or at night. So man had to invent a timepiece, which would be independent of the sun.

A method of doing this was known in Egypt where time was measured by how much water had flowed out of a pot through a small hole. The Egyptians soon discovered that if the sides of the container were vertical, then the flow of water was greater when the vessel was full than when it was empty, and its timekeeping was therefore irregular. These

4 A Tibetan priest's time-stick from Darjeeling nineteenth-century

5 The cast of an Egyptian water-clock from Karnak temple, 1415-1380 BC

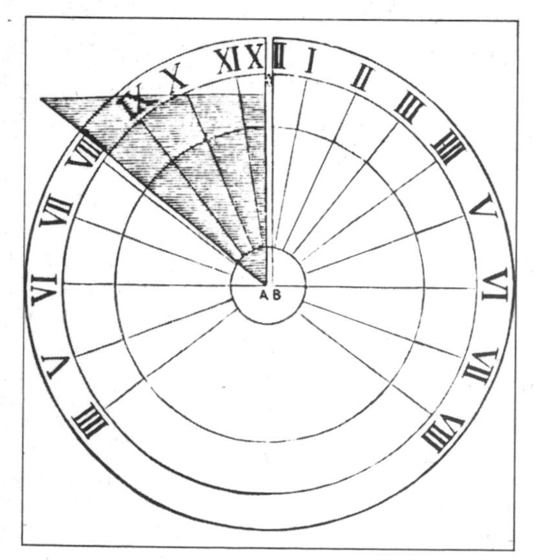

6 Diagram illustrating the working of a sundial

7 A model of an Egyptian shadow-clock, 1000-800 BC

8 Iron fifteenth-century chamber-clock, probably German, with an alarm, height 20 ins. This is now in the Clockmakers' Company Museum, London

scientifically minded people surmounted this difficulty by making the sides of the container at an angle of seventy degrees, then the water level fell at a uniform rate and the time was gauged accurately [figures 5, 10]. Milk was used instead of water in one of the great temples, since the god was known to prefer it. Often these Egyptian water-clocks were intricately decorated (just as, in the eighteenth century in Europe, mechanical clocks were fancifully ornate) and in the Cairo Museum there is a water-clock in the form of an ape urinating. This method of measuring time was also applied in reverse—by how much water flowed into a container rather then out of it. A bronze bowl, with a small hole in the bottom, was placed on the surface of the water which then slowly leaked into it [figure 16]. After a calculated period of time the bowl sank.

9 A plaster cast of a Saxon sundial on Kirkdale Church in Yorkshire

These 'sinking bowls', used in Algeria for centuries, are still employed there to-day to time the amount of water to which farmers are entitled for irrigating their land. Sinking bowls are also used in India, by the Brahmins. Such bowls as these have also been found in Ireland, and they were used by the ancient Britons. The Alexandrian scientist Ctesibius in 124 BC, made a water-clock that looked almost like a weight driven clock, for its face had a round

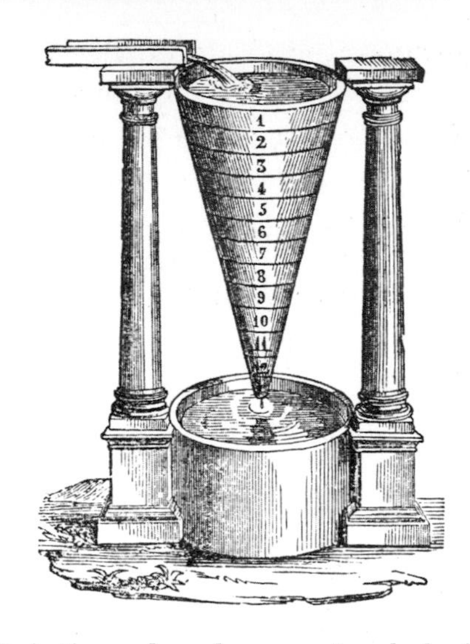

10 Antique *clepsydra* or water-clock similar to those used in Egypt

dial and a hand, but instead of the weight there was a float which rested on the water, and as the level of the water lowered, so the float pulled down a cord which caused the hand to move round. The use of water-clocks was wide-spread not only in Egypt, Syria, Babylon and Phoenicia but all over that part of the world we now call the Middle East. In the first century BC Syria gave a monumental water-clock to the city of Athens [figure 15]. It was an octagonal tower, part of which could be filled with water. On each of its eight sides, there was a sundial, and because it was carved with figures of the winds on each of its eight sides, it was known as the Tower of the Winds. It was the official timekeeper of Athens, and it is still standing today. Small water-clocks were used in the

11 The obelisk sundial erected in Rome by the Emperor Augustus; from an old Italian print

12 A German table clock, made anonymously *c.* 1580. Its gilt dial shows the hour figures I-XII and 13-24 and a central sun. It is now in the Clockmakers' Company Museum

Athenian Courts to limit the time a lawyer was allowed to speak. This led to those who filled the water-clocks being bribed by lawyers to fill only partially the container when rivals were pleading, and so reduce their opponents allotted time, and the same thing used to happen in Rome under Pompey. There is a story that a Roman prisoner's life was saved by a pebble falling into the water-clock, which made the water drip so slowly that it gave extra time for some delayed evidence to arrive, which acquitted him of murder. The Romans were proud of

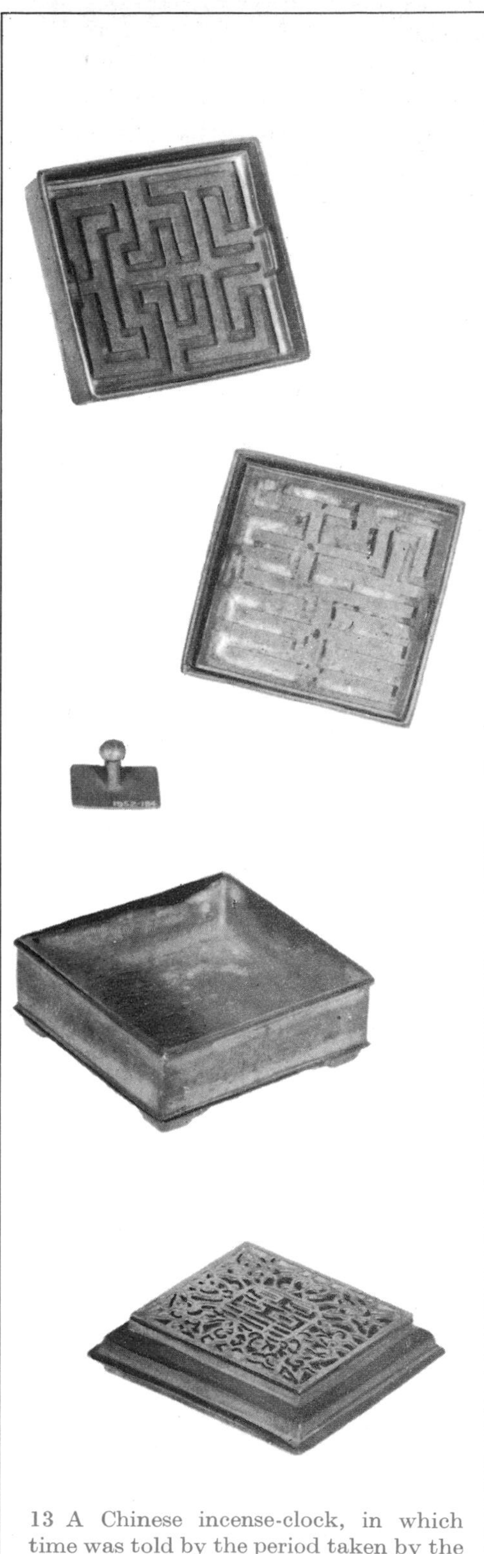

13 A Chinese incense-clock, in which time was told by the period taken by the incense to burn through the maze

14 A fifteenth-century striking chamber-clock, anonymous and probably German, height 15 ins, now in the Clockmakers' Company Museum

their water-clocks, and Vitruvius gives us a detailed drawing of a brilliantly complicated monumental water-clock in Rome in the first century AD. It is said to have come as a surprise to Julius Caesar in his campaign against the Gauls, to find that they too had water-clocks. There is an account in a medieval manuscript of how a fire in a French monastery was swiftly put out with the water from the monastery's clock.

15 The Tower of the Winds, Athens; from an old print

Islam, which drew its cultural inspiration from the old civilisations of the Middle East, added automata to water-clocks, so that small animals and figures moved and birds sang when the clock struck the hour. (The Arabs even invented a clock-work machine that registered a patient's blood pressure.) This invention of automata or mechanical figures took place *c.* 900 AD when the Arabs were at least three centuries ahead of Western Europe. So that when the Caliph of Baghdad presented a water clock with automata to the Emperor Charlemagne at Aix in 809 AD, it made a great effect. At each hour, one of the twelve doors that marked the hours on the dial opened, from which dropped the correct number of little copper balls onto a metal bell beneath. Then at noon, twelve Knights appeared who shut the doors again, and according to a writer of the day, 'the clock had many other astounding features never before seen by our Frenchmen'.

Water-clocks were not unknown in ancient China, where each day was divided into a hundred parts. The recent translation of a long-neglected manuscript written by Su Sung (1020-1101 AD) mentions old Chinese clocks which seem to have been in advance of the other clocks of the same period about which we know. Powered by a water-wheel, Chinese clocks were the first, as far as we know, to feature that mechanism, which prevents the power from running away unchecked, which has become an essential part of all clock making. This mechanism, re-discovered centuries later in Europe, lets the power escape only little by little and is nowadays known as the escapement. This Chinese variety was invented by Chang Sui (682-727 AD), a Tantrist monk, working with Liang Ling -Tsan.

16 A Saxon 'sinking-bowl', a type of water clock

Yet waterclocks suffered from the same disadvantage as sun-clocks—they depended on the elements. In short, water-clocks were apt to freeze up in winter! It was perhaps for this very reason that in the thirteenth century Alphonso X of Castille had a clock in which mercury was

17 An eighteenth-century oil-clock which told the time by wastage, from North Germany

18 A set of sand-glasses in a brass frame

substituted for water, which worked wonderfully well, for mercury freezes at 38° centigrade below zero, and so the clock was never stopped by a frost.

Yet another way of reckoning time was by wastage, that is to say by discovering how long it took a candle, or incense, or oil to burn away [figure 17]. Alfred the Great, King of Wessex, when a fugitive in his own country, vowed that if his kingdom was restored to him, he would spend a third of his time in God's service. This vow he afterwards fulfilled, giving eight hours a day to religion. To measure the time, he had two candles, each twelve inches in length, which burned four hours each (an inch

19 A *bronze doré* globe with a clock movement made by Jost Burgi at Cassel in 1580, now in the Conservatoire des Arts et Métiers, Paris

20 A seventeenth-century striking clock. As the hours are sounded the Negro's head moves and the dog jumps at his feet

in every twenty minutes). So that draughts should not alter the speed of burning, the King had horn scraped thin enough to to be transparent, and set it in lanterns. Thus each candle could burn evenly in all weathers. In the same way St Louis used a candle-clock for measuring the time he spent at his devotions, and King Charles V of France divided his day into four equal parts with a candle-clock. The ancient Chinese sometimes reckoned time by how long it took a specially prepared incense to burn away [figure 13], while another somewhat similar device was a short rod made of sawdust and pitch placed in a small boat-shaped vessel, across which two copper balls were hung on a thread. When one end of the rod was lighted, after a given time the fire reached the thread, and snapped it so that the balls fell with a clang on the metal tray beneath. A rather simple though effective way of telling the time was used in medieval monasteries, when there was no other way of knowing the hour at night. Brother Austin would begin to read his Bible, and, after he had got through a specified number of previously timed pages, he would run to the belfry and ring the bell. He was in fact a monk-clock.

Lastly we come to sand—or hour—glasses, in which time is determined by a quantity of sand passing through a narrow neck between two bulbs of glass [figures 18, 22]. These are supposed by some to have been employed originally on sailing ships; but another insists that they were used by the Roman army to measure the watches in the night, and a third declares them to be the priceless invention of a monk at Chartres, who at the end of the eighth century resuscitated the art of glass blowing and

21 'No, no, no! *Thirty* days hath September'. An Aztec calendar-clock as drawn in the *New Yorker* 1960

invented them. Whatever the truth may be, sand-glasses certainly came into great prominence in the fourteenth century, and were still highly treasured objects at the court of Queen Elizabeth of England two centuries later. In those days they were used to time knightly bouts in tournaments; to-day they stand on our kitchen shelves to tell when our eggs are boiled. There was a monumental sand-glass which stood out of doors in a Berlin park until the last war.

It is important to remember, as we pass on to the age of mechanical clocks, that their coming did not put an abrupt end to the older ways of telling the time. Water-clocks went on being used for many years, while sundials and sand-glasses have gone on for ever. Old ways of knowing the time ran concurrently with the new.

> The natural clock-work of the Mighty One
> Wound up at first and ever since has gone.
>
> *Inscription on a sundial on Seaham Church, Durham, England.*

22 A set of four sand-glasses measuring the quarter-of-an-hour, half, three-quarters, and one hour, made in 1720

Early Domestic Clocks

23 'Time is man's angel'. SCHILLER

WESTERN MAN HAS USED mechanical clocks only for the past seven hundred years. After centuries of telling the time by the methods described in Chapter One, at last man came to measure time by weightdriven clocks. Considering that these work on the same principle as that of the bucket going down the well and forcing the handle round (which had been going on from time immemorial), it is a wonder he did not hit on the idea sooner [figure 27]. But what was to prevent the weights from running down unchecked faster and faster? A device was

24 A striking lantern clock, height 13 ins, signed 'Jeffry Baylie at ye turn-Style in Holburn fecit' and now in the Clockmakers' Company Museum

25 A late fifteenth-century wrought-iron clock from Würzburg

26 A standing clock made by F. L. Berg in Augsburg in 1719, now in the Clockmakers' Company Museum

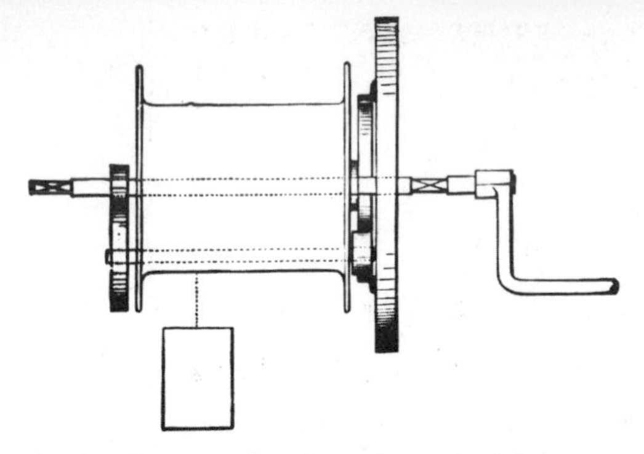

27 A diagram showing the principle of a bucket descending a well, as applied to weight-driven clocks

engineered which allowed the driving force of the weights to escape only bit by bit and with regularity—the revival of an idea, as we saw in the last chapter, incorporated in ancient Chinese clocks, called an escapement. Many varieties were subsequently made. The first of these in Europe was the Verge escapement, an amazingly clever mechanism, impossible to describe here adequately in non-technical language. The early history of mechanical clocks is still uncertain and obscure. It has been suggested

28 The under-side of a table-clock showing its dials, astronomical tables and a compass. It was made by Jacob Marquart, Augsburg, 1567

29 This allegorical figure of *Arithmetica* carries a weight-driven clock; from a sixteenth-century engraving by Marten de Vos

that they were an Arab invention and were brought to Europe by the Crusaders. As the Arabs were far ahead of Europe in scientific attainment at that time, this may very likely be true. In early illuminated manuscripts, the figure of Prudence or wisdom is usually shown holding a weight-driven clock, and so they are traditionally called wisdom clocks. They appear in Flemish tapestries; fine examples can be seen in the National Museum in Madrid and in

30 A drum-shaped table-clock with an alarm

31 A drum-clock with a detachable alarm, *c.* 1600

32 This German clock, made about 1600, is one of the earlier wheel-lock alarm-clocks

33 A gilt spring-driven rack clock built by Mosbrücker at Saverne *c.* 1780. Marketed by Mabille of Paris

the Glasgow Art Gallery. Prudence stands there in very much the same way as the figure of *Arithmetica*, illustrated in these pages [figure 29]. But we really know a great deal about the masterly astronomical clock made by Giovanni de Dondi in 1364 in Italy because he left behind a full description of it. This clearly shows that it could not have been better designed, even with all the advances in mechanical knowledge, to-day; and that it was apparently made of brass, unlike other early clocks, all of which were of iron [figure 25]. This Dondi clock remained in Italy until 1585 when it was taken to Spain and in 1809 was destroyed in the Peninsular war. From his complete drawings which have luckily survived, an entirely new clock has been only recently reconstructed under the supervision of the well-known horologist H. Alan Lloyd, and this is now in the Smithsonian Institute, Washington. The first clocks were made by blacksmiths and locksmiths in an iron frame, and were designed for putting on a wall bracket so that the weights could hang down unhindered [figures 8, 14]. Italy was the brilliant pioneer; but soon the craft was carried northwards to the South German towns which became world famous for clock-making, Nuremberg, Augsburg, Cassel and Ulm. In these cities clockmakers had to obey certain guild rules, assuring standards of perfection that have seldom been surpassed [figure 36]. Before a clockmaker was allowed to practise his craft, he had to make a satisfactory horizontal, square, or hexagonal table-clock (the choice was his) and he was given eight months in which to do it [figure 28].

About 1500, in Italy, France and later in South Germany, springs were being introduced into clocks in place of weights. The invention has been ascribed to Peter Henlein of Nuremberg, but the idea is illustrated in Leonardo da Vinci's note books, though it is not known whether he ever applied it practically. Initially these spring driven clocks were shaped like a drum, a few inches high and perhaps six inches in diameter [figure 12]. The face was on the flat top, and sometimes protected by a metal cover pierced with a design to enable one to see the dial. A development of these drum-shaped table clocks was the addition of a detachable alarm mechanism [figures 30, 31]. Another sort of alarm clock, one of the first of the table variety ever made, was the German wheel-lock type [figure 32]. This clock was set for whatever hour an alarm was wanted, and a wheel-lock, automatically set off,

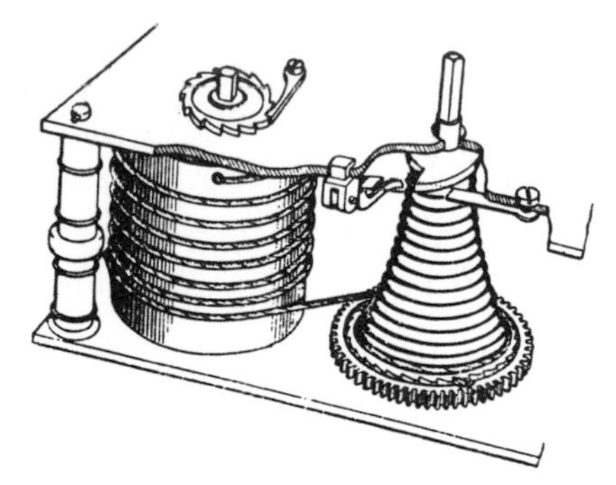

34 The *fusee* assures a uniform drive from a spring whose power decreases as it uncoils

35 A standing table-clock, height 14½ ins, made in Augsburg in 1506

produced a spark which ignited a gunpowder charge, and at the same time, lit a candle. Far later, about 1832, the idea was applied to a sundial in the Luxembourg gardens in Paris. A burning-glass, fixed at the mid-day position shown by the sundial, was focussed on a charge in a cannon, and exploded accordingly. These were the forerunners of those convenient alarm devices we enjoy today, such as the one which raises your bed from the horizontal to the vertical position and the one that makes you a cup of tea and switches on the radio.

Eventually, clocks of upright form began to be made in Augsburg and copied elsewhere [figure 20]. Many were fitted with elaborate astrolabe and calendar dials, incorporating complicated gearing apparatus [figures 35, 37]. Astronomy and clocks go together because time, in its broadest aspect, is measured by the movement of our earth through the sky. Domestic astronomical clocks were not only treasured by professional and amateur scientists, but also by those who believed in the fortunes of the stars. If one knew what sign of the Zodiac one was born under, one could see from an astronomic clock the position of the planets and stars and thus gauge whether the time was propitious or not. One of the greatest makers, Jost Burgi, was born in Switzerland in 1552, and when he was working at Cassel in Germany, improved the time standard of his day considerably by devising his cross-beat escapement which successfully modified the verge escapement. It can be seen in his astronomical clock which he made in 1580, now in the Conservatoire des Arts et Métiers, Paris [figure 19].

Because spring-driven clocks are so compact (they can be put in any position and still keep going) their exteriors could become fantastic—a blend of fact (time) and fantasy (the artistic design). Human and animal robot or mechanical figures gesticulated on them [figure 38]. Domestic automatic clocks were now the most delightful of diversions, and guests were entertained with them [figure 26]. In this mood, Hans Schlotheim made a celebrated 'Tower of Babel' automaton clock with a rolling ball that 'fell down' to music, as well as the golden sailing-ship clock, which at certain hours, pitched and rolled. A. Kirchen designed a sundial as a flower, a conceit since copied and made into a clock by an enterprising firm in Switzerland. Much less complicated, and for owners who wanted the time without embellishments, was the typical

36 A Gothic brass-gilt clock made in South Germany *c.* 1520

37 A gilt-bronze table-clock made in Augsburg, *c.* 1560

38 This gilt-copper clock has a silver dial supported by a griffin whose beak and wings move when the hour strikes. It was made in South Germany, *c.* 1620

domestic clock of the early seventeenth century, the so-called 'lantern clock' made entirely of brass [figure 24].

At about the same time as the craft was spreading northwards, it moved west across France to Blois. Francis I had invited Italian Renaissance craftsmen to his palace and with them they brought clocks. These the French interpreted in their own special way, adding a brilliant striking mechanism much ahead of that of their competitors, which is still in manufacture [figure 39]. Francis I granted a charter to the Paris Clockmakers Guild. A portrait of Henry II, his successor, with a typical French drum-table clock of the period, adorns the Musée Condée at Chantilly.

Because a spring pulls harder when tightly wound up than when it is run down, ways were sought to make it pull more evenly, and its timekeeping therefore more precise. The French from the first used the *fusee*, which furnishes an almost perfect solution to the problem of getting a uniform drive from a spring whose force decreases as it uncoils [figure 34]; while the earliest clock in existence to possess a *fusee* was made by the Czech Jacob, who is thought by some to have invented it. The *fusee* is shaped like a truncated cone with spiral grooves which hold a cord, chain or gut, in place round it. This cord is joined to the barrel of the main spring. When tightly wound up the spring turns the *fusee* by pulling the cord from the smallest diameter of the *fusee*; when it becomes wound down, it then has to pull the cord from the larger diameter of the *fusee*. In this way, by pre-adjusting the diameter of the *fusee* to the varying pull of each turn of the spring, the

39 A French gilt-metal clock of the sixteenth century

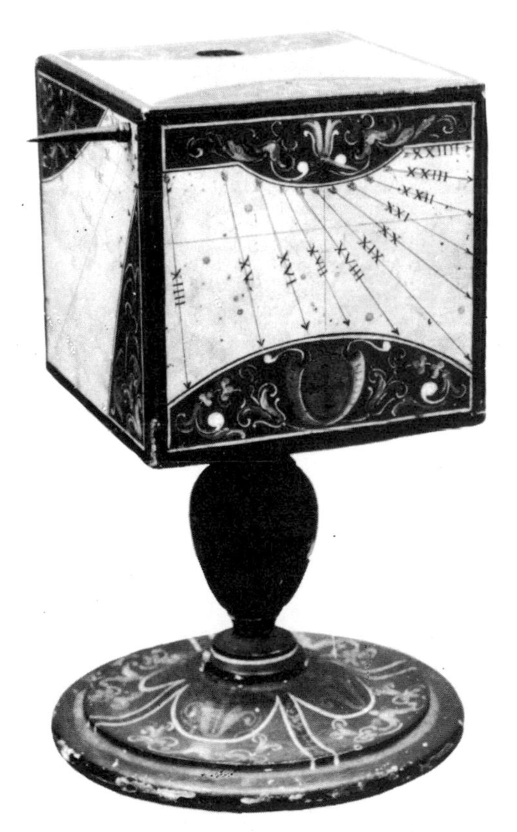

40 A Florentine cube-shaped sundial, *c.* 1560, showing the Italian hours

41 (*right*) *Le Songe de la Vie* by Antonio Pereda [1608-78] which illustrates the theme of time's passing

42 Sand-glass which belonged to Mary Queen of Scots

43 The interior of this German chalice-sundial shows the hours; made in 1596

force on the mainwheels of the clockwork is uniform.

Medieval clocks were luxuries, not the necessities they are now. Having to be 'on time' to the minute for an appointment was not absolutely necessary until the railways and the demands of commerce made it so. Clocks were status symbols. Kings and princes and prelates had them and ambassadors used them both for themselves and as diplomatic presents. But these early clocks, though a source of pride to their owners, were not particularly accurate. Tycho Brahé the Danish astronomer wrote: 'I have been disappointed in putting my clocks in for accurate work.' The ordinary sundials and sand-glasses were more to be trusted [figures 40, 42, 43]. Nevertheless, clockmaker's corporations had already been founded in Paris (1544), at Nuremberg (1565), Ansbach (1591), Blois (1600), Geneva (1601). Among the statutes of the Annaburg Corporation (1605) there are some most curious regulations; masters had to give their apprentices bath money once a fortnight. It one member hit another, he was fined a week's pay, but if he went to hit him and missed, he was fined a fortnight's pay.

Increased interest in clocks dramatized in people's minds the consequences of Time passing, which fascinated them; and brought about a demand for paintings composed of spent objects such as an emptied sand-glass, a skull, a burnt-out candle and a falling-petal rose. Jan 'Velvet' Breughel excelled at this, as did the mid-seventeenth century artist Antonio Pereda [figure 41]. The most important discoveries in clock-making, as we shall realise, have been the use of weights and springs, the fusee and various escapements, the pendulum, the application of electricity, and of quartz crystals and of atomic vibrations.

It was customary for each clock centre to benefit by copying the mechanical discoveries of the others, while exterior decoration followed the fashions in architecture and furniture at the time. To horologists a good clock will very likely mean that the cog-wheels are put together with fascinating ingenuity; to a child it means a living creature with innumerable changing expressions; but to most of us the phrase 'a good clock' means that the case is well made, and that it keeps excellent time.

In creating mechanical clocks, man began the entire machine age. Today he is perhaps the victim of his own medieval cleverness, for the reason that modern machines insist that he shall always be on time, and he has to obey.

Domestic Clocks 1500-1700

44 This seventeenth-century French table-clock is supported by Hercules

There was a man, he had a clock, his name was
Mister Mears,
And every night, he wound that clock for five and
forty years.
And when at last that clock turned out an eight-day
clock to be,
A madder man than Mister Mears, I never hope to see.

TRADITIONAL

UNTIL THE MIDDLE OF THE seventeenth century, clocks were not at all accurate. For this reason they seldom had more than one hand, which indicated the hour only. And their time was continually being checked on a sundial. The invention that made all the difference to accuracy was the pendulum. It has a curious history.

In 1583 Galileo Galilei, a student of eighteen, when watching the to-and-fro swing of a lamp on a long chain in Pisa Cathedral, thought that its movement might be a way of measuring time [figure 45]. From this idea he attempted to make a pendulum clock, but we are not sure that he succeeded.

However, in 1657 Christian Huygens, a Dutch scientist, [figure 46] realized Galileo's dream, and revolutionized the clock-maker's art by successfully applying the pendulum to clockwork. In this way accuracy was so much improved that minute hands became the rule, and soon second pointers were added. Huygens designed these pendulum clocks, and Samuel Coster made them. Both the designs and clocks can be seen at Leyden Museum. Thirteen years later a considerable further step forward in accuracy came with the invention of the 'anchor' escapement, often attributed to Robert Hook but first met with in clocks by William Clement, a London maker of about 1670. It got its name from the fact that the gadgets which control the escape of the driving-force were anchor-shaped. In America the various parts of the anchor escapement are to this day known by their antiquated old English names, whereas in Europe the old names have been changed.

45 A model showing Galileo's suggested application of the pendulum to clock-making

These two things, the pendulum and the 'anchor' escapement, had an immediate effect on clock manufacture. The greater accuracy now attainable made it worth while to contrive mechanism that would go for a longer period than thirty hours which had been the limit before. Soon clocks were devised to go for a month or more, then for three, six or twelve months. This meant that the mechanism required much heavier weights which unfortunately pulled the bracket holding the clock from the wall; so a tall, upright, coffiin-like case was provided. It not only strongly supported the clock but enclosed the weights as well. Thus the grandfather clock was evolved, which, when less than six feet tall, is known popularly as the grandmother. At first these long-case clocks were slender, because the case only needed to be wide enough to contain the weights, as the clock had a very short pendulum. But after Robert Hook had shown by experiment that a long pendulum worked better than a short one, then grandfather clocks had to be made wide enough for the swing of the pendulum as well. A whole new branch of cabinet making was called into being to create elaborate long clock cases, some of marquetry [figure 51].

In Friesland and Zaandam in Holland a more ordinary kind of weight-driven wall clock was made with a hood [figure 52]. This Friesland design is the link between wall clocks and grandfather clocks. In 1660 wood carvers in the Black Forest managed to make clockworks entirely of wood. This discovery marked the beginning of a large industry based on the town of Furtwangen. These 'Deutsch' (frequently misnamed Dutch) clocks, which later had brass works, were to be found in almost every house until cheap American clocks came to Europe.

46 Christian Huygens [1629-93] who first successfully applied the pendulum to clockwork. A portrait by an unknown artist

In the same year Charles II was restored to the English throne, and the wonderful array of continental ideas and knowledge this king brought in his train when he returned to London stirred up Englishmen's minds and made them 'active, industrious and inquisitive' (Sir Thomas Sprat). This applied to clock-making particularly. In 1631, a royal charter had been granted to London clockmakers, its rules applying to the territory ten miles around London, and giving the Clockmakers' Company the power to enter any premises and to destroy bad work. At the Restoration it was Ahasueras Fromanteel, assistant to Samuel Coster, Huygens' clockmaker, who first brought pendulum clocks to London [figure 50]. This was the great age of clock-

47 A late seventeenth-century Inclined Plane clock, which goes by its own weight. When the clock has finished its descent, it is lifted off and placed again at the highest point

making in England. Tompion, Quare, Knibb and William Clement were amongst the great interpreters and inventors [figure 49]. The twenty-four hour system of measuring time which, owing to the sudden great increase in clock production everywhere, had been, with one or two exceptions, generally adopted across the western world, had entailed the imagining by mankind of an average or mean sun, which unlike the real sun, had hours of exactly the same length. Thus its imagined position would fit in exactly with the evenly spaced hour marks on the clock dial and the regular character of clockwork in general. That is what it amounted to and it has worked very well ever since. But to many a rustic mind the idea of an average sun naturally seemed to be nothing but a

48 A falling-ball clock, 4 ins in diameter inscribed 'Jacob Behan, Vienna'

49 A clockface by Thomas Tompion, one of the more famous of English clockmakers, made for Sir Jonas Moore, 1676

townsman's hoax. In a village near Chester in the West of England a clockmaker, Peter Clare, was derided when he asserted that his mean time clocks were right and the sun was wrong. There, a memorial reads:

Here's the cottage of Peter that cunning old fox,
Who kept the sun right by the time of his clocks.

To keep clocks right to-day we can use radio and television signals, but in those days you checked them with sundial plus equation tables. This latter was a chart showing the difference at all hours between true solar time and mean solar time. From it, after looking at a sundial, it was easy to work out the correct time to set your clock at. But soon a clock was marketed with a small equation dial which was marked on the right with 'Sun faster', on the left with 'Sun slower', and which had a pointer indicating the exact variation between the two times. There is a perfect example of this kind of clock made by Thomas Tompion in the Pump Room at Bath.

Concurrently a high precision timepiece was made to give the correct time to clocks in Observatories. These were named Regulator clocks.

It was the fact that at this date both the invention of the pendulum and the anchor escapement were available to be applied to the clocks which Thomas Tompion designed and made for Greenwich Observatory [figure 49] that assured the superiority of the royal astronomer John Flamstead's calculations of time.

Concurrently in France a fascinating character, Grollier de Servière, after spending his youth in the army, concentrated on measuring time by how long it took a small metal ball to run repeatedly from A to B. The results were clocks that looked not unlike some penny-in-the-slot machines that had been elegantly and richly made. He also turned a shallow bowl of water into a clock by floating a small cork tortoise on it and by marking the hours in Roman numerals round the rim. The tortoise swam round the edge from number to number as the hours demanded, following a magnet which moved underneath the bowl. On the same lines he made a toy lizard creeping up a column tell the time, and a mouse moving along a cornice do the same. There seemed to be no limit to the things that could be done by clockwork to make time-keeping amusing and different. A number of other artifacts made their appearance in France towards the end of this century, which were known as gravity clocks; one which was

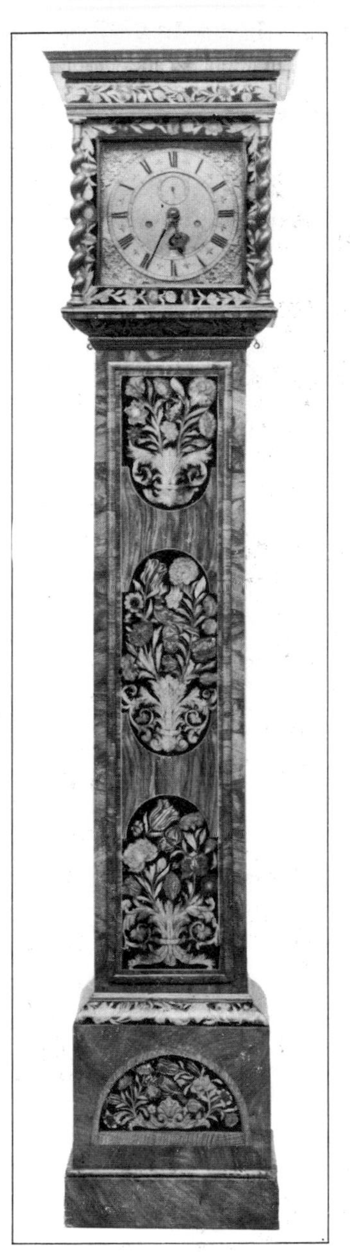

50 (*above*) An English brass lantern-clock by Ahasueras Fromanteel, 1670, height 14½ ins

51 (*centre*) An English long-case clock, *c.* 1680, signed 'Wm Clement Londini fecit'. Its case is of walnut veneer with inlaid marquetry of various woods

52 (*right*) A Friesland clock, said by some to be the predecessor of the grandfather clock

driven by the force of its own weight as it gradually slid down a toothed shaft, was called the Rack clock [figure 33]. This had originated in the late sixteenth century in Germany. Another, called the inclined-plane clock [figure 47] was one that moved slowly and imperceptibly down a slope gathering driving-power from its progress. Yet another, a decorative falling-ball clock [figure 48] descended a cord from the ceiling and showed the time on a horizontal dial-band round its circumference. These mechanical fancies were but a foretaste of the ingenious clocks which were later evolved in the eighteenth century.

French Domestic Clocks 1700-1800

I see that time divided is never long,
and that regularity abridges all things.
MADAME DE STAEL

53 (*left*) A late eighteenth-century French cartel clock, mounted in ormolu. The numerals on the face are in blue enamel, often the practice with ormolu work. From the Victoria and Albert Museum

THE BEST CLOCKMAKERS in France were given lodgings in the Louvre by the King, and considered as artists. They dined in the palace at the table of the Gentlemen of the Chamber and had the right of entry to the King's presence, along with distinguished members of the Household. Every morning when the King dressed, the clockmaker on duty wound up the Royal clocks and properly adjusted the fob-watch that the King was about to wear.

In 1712 Louis XIV had Jeremy and Louis Martinot, Augustus Bidault and Charles Boulle living in his clockmaker's lodgings. Decorative art in France had undergone a remarkable change; the shape of clocks had changed from lines recalling those of religious buildings to the glorious classical forms dictated by eminent and enthusiastic cabinet makers [figure 60]. Pre-eminent among them was Charles Boulle, whose dramatic use of brass, tortoiseshell and turtle-shell inlay quickly made him famous and whose work became the fashion [figure 57]. Windsor Castle has a typical Boulle clock in the Council Chamber and there is a beautiful painting of one in Watteau's picture *L'enseigne de Gersaint*, now in the State Museum, Berlin.

There was an extravagant demand for sumptuous furniture of all kinds [figure 55]. Designs by Marot, Thuret, Caffiéri and Berain kept pace with it. Animal clocks, obviously inspired by the seventeenth-century Augsburg variety, were now made more splendid; horses, lions and camels were introduced as clock carriers. The elephant with a clock on its back is signed '*Fait par Caffiéri*' while the movement is by Jerome Martinot [figure 65]. Its rich ness and elegance are characteristic of the clocks of this period.

To flatter Louis XIV, the clockmaker Burdeau created somewhat out-moded automatic gilt-brass clocks, showing the King enthroned. At each hour, the Electors of the German States and the Princes of Italy moved up to *Le*

54 The heading on the stationery of an eighteenth-century French clockmaker

55 A French gilt-metal clock, *c.* 1700, height 3 ft 4 ins. This clock was restored by Vulliamy in 1854

Grand Monarch and bowed, after which they chimed the quarter hours with their canes. The hours themselves, however, were struck by the kings of Europe, but only after they, too, had paid their respects to the monarch. One of the kings, William III of England, had, in fact, been made by the clockmaker to bow especially low to King Louis, because of his well-known opposition to him. This contraption had a great success at Court. When it was shown to the public, however, the unexpected happened. Some part of the machinery snapped and the French King lay prostrate at the English King's feet. As a result of this disaster, poor Burdeau was confined for the night to the Bastille. Because, during the latter part of his reign, flatterers likened Louis XIV to the sun, pendulums were often decorated with the face of Phoebus, the Sun God. At the same time the intense desire for fresh forms led to the production of cartel clocks (the word cartel probably being derived from the Italian, *cartello*, meaning a bracket). These were either fixed against a wall or on a ceiling and were either oval or round in shape [figure 53]. The popularity of these clocks continued into the next reign [figure 56]. François Boucher shows what they were like in his painting *Le Déjeuner*. In contrast to the French cartel design, in Italy Piranesi designed cartel clock cases based on the shape of an antique Roman ship's rudder.

Exquisitely made long-case clocks throughout the century reflected the fashion of furniture [figures 58, 64]; and these are to be seen nowadays in all the great museums and collections; outstandingly perhaps in the Wallace Collection in London. Possibly the most dazzling personality amongst Louis the XV's clock-makers was Pierre Auguste Caron, for not only did he win an award from the Paris Academy of Sciences for mechanical invention, but he was also an accomplished musician, and the successful author of *The Barber of Seville* and *The Marriage of Figaro* [figure 62]. He is better known under his other name of Beaumarchais. A rival clockmaker, Jean André Lepaute, who was also an *Horloger du Roi*, produced some clocks that were wound by a draught of air going up the chimney (a method which was later re-invented). An amusing jotting from an account book belonging to a third *Horloger du Roi*, Pinon, describing work done for the Comte D'Artois, reads: 'For repairing a movement of a clock in the Prince's apartment, and renewing, in fine bronze, the female figure

56 (*right*) A Louis XVI ormolu cartel clock

57 The case of this clock is by Charles Boulle, and is inlaid with brass, tortoiseshell and mother-of-pearl, and is decorated with ormolu. The mechanism is signed 'Salles à Caen, 1700'

58 (*far left*) A French long-case clock of 1780, signed 'R. Robin à Paris'. Its case, with a panel of inlay in tulipwood and rosewood and surmounted by a gilt-brass figure of Father Time, is by B. Lieutard. It is now in the Victoria and Albert Museum

59 (*left*) French mid-eighteenth-century animal clock. The ormolu case, with a frame of brilliants for the dial, is carried by a silver horse

60 An early eighteenth-century bracket clock, signed 'Bastien à Paris'. Its case is decorated with an inlay of brass and tortoiseshell and with ormolu mounts; now in the Victoria and Albert Museum

61 A Louis XV ormolu and bronze mantel-clock of the Rococo period, signed 'Benoist Gerard, Paris'

on the clock-case, which the Prince had amused himself in scratching with a knife from one end to the other, with the object of cleaning it, renewing the clock and other accessories, Livres 2,068'. A fourth Royal clock-maker, Jean Antoine Lepine, came to Paris from Switzerland when he was twenty-four, and became a friend of Voltaire, who started a factory at Ferney to help forty clockmakers who had been exiled from Geneva, with whom he was in sympathy. Alas, the factory enjoyed success only as long as Voltaire was able to persuade his friends to patronise it.

62 Caron de Beaumarchais, clockmaker and playwright, by an unknown artist

It could take twenty years to calculate the various movements for a really complicated astronomical clock and twelve years to construct it. That was how long it took Admiral Passement to work out the exact specifications and Danthiou, under his directions, to build it for presentation to the King at Versailles in 1750 [figure 63], and this happily is still at the palace.

But greater than all these were Pierre Leroy, Ferdinand Berthaud and Abraham Breguet. Leroy inherited genius from his father, Julien, who had himself invented a clever repeating mechanism of the first order. Pierre invented a brilliant chronometer device. Berthoud, born in Switzerland in 1727, settled in Paris when he was nineteen. The books he later wrote claimed to contain all that was then known about clocks. A regulator clock of his is said to have been taken from the Tuileries in 1793, after being white washed to hide its value. If this is in fact so, then it must be the regulator clock by him now in the Wallace Collection. Abraham Breguet had great gifts, for everything he did worked unusually well and yet bore his highly original inventive stamp. Perhaps his most important feat was to make a clock that was able to regulate a watch [figure 67]. It would correct the watch providing it was not more than twenty minutes out. It worked like this: at an appointed hour a plunger was released from the clock which engaged a pusher in the band of the watch. In its turn the pusher contacted a wheeled shaft which set the minute hand to time, and moved the regulator in the appropriate direction to compensate for the error; while at the same moment another plunger from the clock wound the watch.

Mantel clocks before the reign of Louis XV were exceptional, but came into their own when the fashion for Rococo design started, and clock cases, cartel, mantle and all the rest took their decorative effects from shells and luxuriant vegetation arranged to make a fanciful confusion of arabesques in which symmetry was purposely avoided [figure 61]. The little human figures on these clocks were sometimes copied from the pictures of Pillement.

Towards the end of Louis XV's reign, the Rococo style had degenerated into a riotous travesty of the original. Reaction to the Rococo rampagings brought about the simpler, more balance designs of Louis XVI [figure 68]. A foretaste of this move towards simplicity can be noticed in the clock beautifully painted in the picture *Reading of*

63 An astronomical clock by Passemont and Danthiou, built for Louis XVI at Versailles

64 (*left*) An elegant French provincial long-case clock *c.* 1770, here photographed at Hamish Antiques Ltd

65 A mid-eighteenth-century clock by Jerome Martinot of Paris. Its case of chased ormolu, carried by a bronze elephant on an ebony and ormolu stand, is signed 'Fait par Caffiéri'. From the Victoria and Albert Museum

Molière by Jean François de Troy (1679-1752). It is an excellent mantel clock with Father Time bearing the dial and works on his back, giving the clock face the prominence it should always have, and didn't always get. It would be hard to find a more elegant clock than the lyre-shaped time pieces that were made in this reign. One such, signed 'Kinable', is supposed to have belonged to Marie Antoinette. The case is made of blue Sèvres porcelain and is two feet high. The lyre strings are the upper part of the

66 A drawing of a fantastic bird-cage clock made in 1780 by Swiss and French craftsmen. At every hour the birds sing and a fountain plays

67 Breguet's 'synchroniser', a combined clock and watch, the watch being set to time by the clock. Made in 1814 for the Prince Regent, it is still in the Royal Collection

68 A vase-shaped clock of Sèvres porcelain mounted in ormolu, made in the reign of Louis XVI

pendulum, the lower part ends in a circle of brilliants surrounding the dial. The pendulum moves to and fro on either side of it [figure 69].

The eighteenth century in France was fertile ground for stylish, exotic clockmaking [figure 70], at times even eccentric [figure 66]. Perhaps the best known of these eccentricities is the bronze head of a Negro woman which showed the hours in one eye, the minutes in the other. Marie Antoinette's model of this is now the property of the Comte de Ribes, Paris. There is another similar one in the Metropolitan Museum, New York. A third in a slightly different style was exhibited fairly recently at the Conservatoire des Arts et Métiers, Paris. A clock case designed in this way but minus its works is in the Goelet Collection. A beautiful example by Lepine and Vulliamy is functioning precisely at Buckingham Palace.

69 A late eighteenth-century lyre-shaped clock, signed 'Kinable' and said to have belonged to Queen Marie Antoinette. The case is of Sèvres porcelain with ormolu mounts; the dial is of painted enamel; and the pendulum is decorated with paste brilliants

70 Another beautiful clock which may have been made for Marie Antoinette. Its case is made of Sèvres porcelain mounted in chased ormolu. It is now in the Victoria and Albert Museum

European Domestic Clocks 1700-1800

71 A tortoiseshell and ormolu clock made by Edward Prior of London for export to Turkey. The dial has Turkish numerals

FRENCH CLOCKMAKERS, HOWEVER, were not the only craftsmen to enjoy royal patronage. As we have seen, a court inevitably attracted all kinds of fine workmanship, in furniture, fashions and all the decorative arts, and this was true not only of the French court at Versailles, but at those of nearly all the other European monarchies.

Looking back for a moment, Henry VIII of England bought clocks from the Nuremberg workshops. Mary Queen of Scots owned a fascinating clock-watch in the shape of a human skull [figure 72]; while her cousin, Elizabeth of England, employed Bartholomew Newsam and Nicolas Urseau as her clockmakers. The Empress Catherine of Russia offered the English clockmaker John Arnold a thousand pounds for a clock so small that it could be worn as a ring, a copy of one he had made and sold to clock-loving George III for five hundred pounds. Arnold refused. George III also paid £1,178 to the ingenious Scottish clockmaker, Alexander Cumming, for a clock which registered the height of the barometer during each day throughout the year; and gave him a yearly salary of two hundred pounds to keep it exact. Cumming was one of the most inventive clockmaking mechanics at that time. The Prince Regent patronised James Tregent, clockmaker at 35, The Strand, who was a Frenchman, working in London. He was a friend of the actor, David Garrick. Now Garrick, when rehearsing at Covent Garden Theatre, used to complain of the playwright Sheridan's unpunctuality so often that eventually, together with the manager, he gave Sheridan one of Tregent's striking clocks, which had an alarm movement. Unfortunately history does not relate whether or not Sheridan became more punctual. In 1835 King William IV of England asked Benjamin Lewis Vulliamy, his clockmaker, to alter and adapt a clock originally made for the Queen's Wing in St James's Palace, so that it could be put at Hampton Court. This was done, and only last year it was given a newly gilded dial. Mention must be made, at this period, of the Emperor of China's patronage of the London maker James Cox, who made automata for Ch'ien Lung's Imperial Palaces at Peking. One of these automata clocks

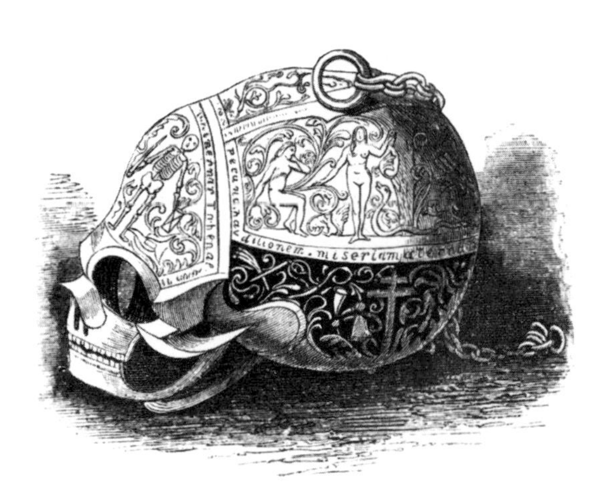

72 The watch-clock belonging to Mary Queen of Scots. The top of the skull lifted to reveal a dial

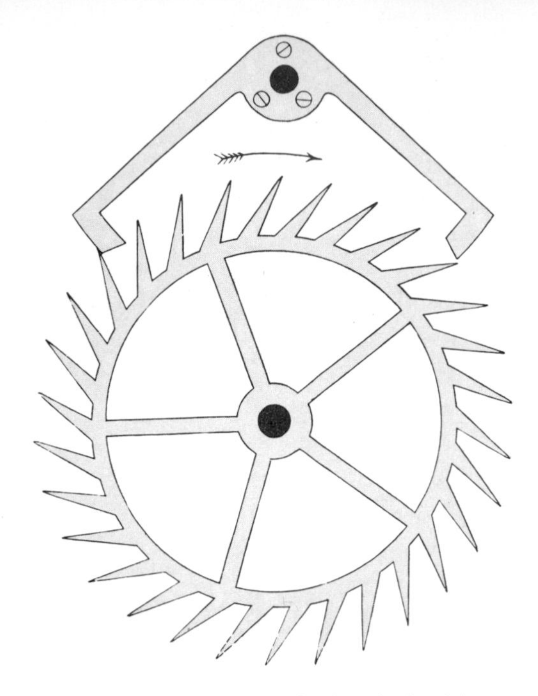

73 A diagram of George Graham's dead-beat escapement

74 The famous Farnese Planisphere

had a crystal dial on each side so that the workings could be seen. Its gold case was said to have been decorated with a hundred thousand precious stones; emeralds, rubies, diamonds and pearls. These are only a few instances among many of how horology has thrived under royal patronage.

The aristocracy followed the fashion set by their princes. Dorothea Sophia, of the aristocratic House of Farnese, in 1725 purchased what is probably the most important astronomic domestic clock ever made from Bernardo Facini of Parma, which was designed by the mathematician Montanari. It has become known as the Farnese Planisphere [figure 74]. Roughly two feet high and going for four days, it was presented to Pope Leo XIII by the Count of Caserta. By 1903 it had fallen into disrepair, so the Holy Father sent it to the Roman clockmakers Hausmann et Cie, who miraculously repaired it. Each piece had to be re-designed, and the missing parts reconstructed. The blue-prints for the job occupied twenty-four drawing tables. It was set going after thirteen months of work, and now it is in perfect running order in the Vatican Museum.

It was in the Black Forest in the middle of the seventeenth century that the first cuckoo-clock sang; one of the first is now in the Furtwangen Museum. It has a square face with a semi-circular pediment. The kind of cuckoo clock we see nowadays, with a gabled roof surmounted by a stag's head did not come in until a century later. This actual design was based on the station master's office at Furtwangen. Some cuckoo clocks were once taken by missionaries to the South Seas where the natives believed that each clock had a spirit inside it which detected any thefts from the missionaries. The High Priest, however, loved these cuckoos, and accepted one as a peace offering. No sooner had he got it home than he took it to pieces, hoping to discover its secret. When he could not reassemble it, he was dismayed, but when the missionaries could not put it together either, he was full of contempt, and had no more time for his visitors.

During a decade of comparatively little royal patronage in England, probably the most important invention during the early eighteeneth-century was made. This was the deadbeat escapement, invented by George Graham in 1715, which for precision purposes remained unsurpassed for the next hundred and fifty years [figure 73]. It gives a slight push to the pendulum almost at its zero position to main-

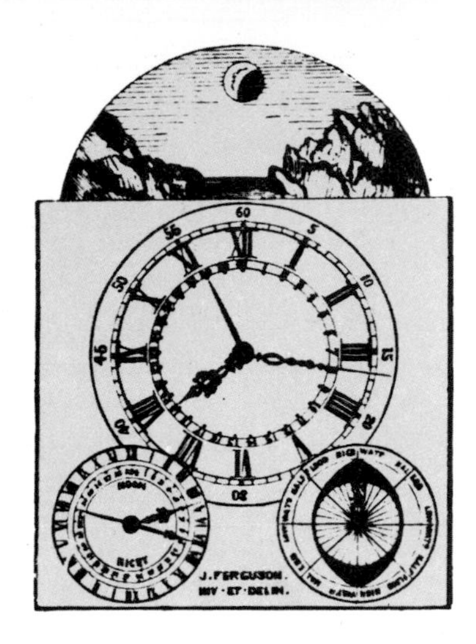

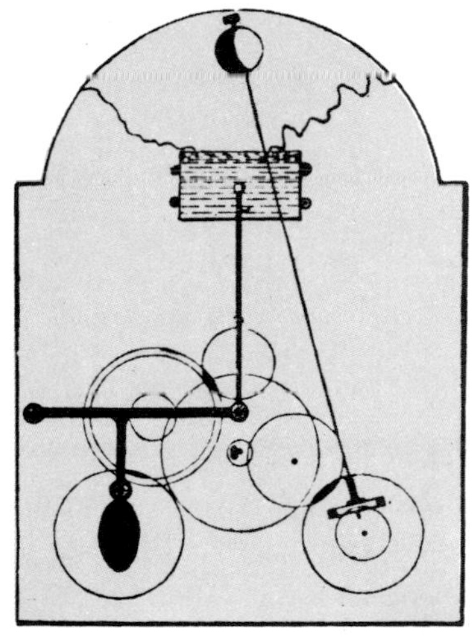

75 A diagram of John Ferguson's tide clock

76 Henry Bridges' 'Microcosm' clock. Its works were checked by Sir Isaac Newton

tain its swing in such a way as to offer less interference to its movement than previous escapements had, thus giving greater exactitude. The fame which English clockmaking enjoyed in Europe during the eighteenth century was to some extent due to George Graham's friendly attitude to his continental competitors. It was against his nature to monopolize his valuable discoveries; so that, for instance, when Julien Le Roy wrote and asked for a description of

77 An American long-case clock of mahogany with satinwood inlay, made in the late eighteenth century

his cylinder escapement in 1728, Graham sent him one. About this time too Arnold Finchett of London produced a fascinating water-clock [figure 82] designed on the same principles as Alphonso X of Castille's thirteenth-century mercury clock. Whilst amongst the clocks made with supplementary dials showing the date, and the phases of the moon, was added yet another variety which gave the time of high and low water and the state of the tides at any given moment at certain ports [figure 75].

It has been thought that British eighteenth-century clockmakers were better appreciated abroad than at home. This is certainly true of Henry Bridges, whose clock, 'The Microcosm' or 'Little World', created more interest among horologists on the Continent than in the United Kingdom [figure 76]. It was ten feet high. On the top a scene showing the Muses on Parnassus was acted, which changed periodically to Orpheus charming the wild beasts in a forest; the middle section was taken up with an array of dials which showed astronomical movements. Since, as we have already seen, these automata stopped at nothing to make an effect, the clock contained an organ which played at certain intervals. Altogether there were said to be a thousand wheels and pinions in the mechanism. At Liège in France, an English Jesuit, Thomas Hildeyard, built a marvel of a clock topped with a crystal sphere, displaying the whole system of the Universe (minus only a few planets which have been discovered since his day). It is now in a Madrid museum. A Spanish catalogue described it as by an anonymous maker but recently a clock expert, Philip Coole, found a print of Hildeyard's clock and identified it. An important influence on European and American clockmaking was had, not only by actual clockmakers like Henry Bridges, George Graham and James Cox, but also by the great English cabinet-makers of the time, Chippendale, Mayhew and Sheraton. They designed grandfather clock cases of a distinction that has seldom been surpassed, and inspired many a designer of elegant long-cases in America [figure 77].

The oldest long-case clock in the United States is signed John Fromanteel and is the Philadelphia Library. The Old Hall clock in the same city bears the name of Devereux Bowley, Lombard Street, London, and there is another in Washington, with the same signature, differently spelt, 'Devereux Bowly'. He was a well known maker of repeating clocks and a member of the Society of Friends, who

78 The celebrated Peacock clock, made by James Cox, now in the Hermitage Museum, Leningrad

79 James Cox's 'Perpetual Motion' clock

bequeathed a large sum to their school in Clerkenwell and five hundred pounds to the Clockmakers' Company, of which he was Master in 1759. London clockmakers were certainly in demand abroad. At the other side of the world, in Russia, the powerful Prince Potemkin bought the Peacock clock from James Cox [figure 78]. Eventually this clock came into the possession of the Empress Catherine; it is now in the Hermitage Museum in Leningrad where its complicated mechanism has been put in running order by expert Russian mechanics. The peacock, large as life, sits on an oak tree stump, shakes its head, displays its tail and turns slowly round. Its movements are remarkably lifelike. On a lower branch sits an owl in a cage decorated with bells; it moves its head and flashes its eyes as the cage revolves, and the bells play a pretty tinkling tune. On a branch near the ground perches a squirrel; on the other side a cock preens and crows. Gourds, ivy and mushrooms grow at the foot of the gilt-bronze tree stump. The top of one of the mushrooms has two small openings through which the Roman figures of the hours and the Arabic numerals of the minutes can be seen. Beside it a cricket moves sixty times in each minute. The pedestal of the entire clock is edged with cut crystal, which gires it a very brilliant appearance.

The clock was sent to Russia from Cox's workshops in pieces, which were packed in several cases. Potemkin invited Ivan Kulubin, the celebrated eighteenth-century Russian automata maker, to assemble it. Kulubin took all the parts to his workroom and laid them out on the floor, and simply contemplated them for three weeks. Finally he noticed that one feather of the peacock's tail was different from the rest and that this was the key to the whole problem of assembling the clock.

Kulubin, however, was obliged to be absent from his workshop for several days for some reason. During his absence a fire broke out in some shipyards nearby, so his son hastily gathered up the pieces, threw them all back into their packing cases higgledy-piggledy, and carried them to safety. His unfortunate parent had to arrange them correctly all over again.

With export markets like this and with the prospect of Central India being opened to trade, it is not surprising that James Cox laid in a stock of extravagantly decorated automatic clocks, made of the richest materials, intent on selling them to the Indian princes. Unfortunately this

80 A regulator clock-movement by Benjamin Vulliamy, 1780

idea came to nothing when further outbreaks of fighting in India prevented export. As these ornate clocks were quite unsaleable in Europe, Cox opened an exhibition of them in a museum at Spring Gardens in London. Half a guinea was the cost of entering this museum. Only a few persons were admitted at a time and the strictest precautions were taken to guard his fifty-six jewel-encrusted automatic clocks and toys, which were

81 A late eighteenth-century English gilt-brass clock, with mechanical figures and a musical box, signed 'Morris Tobias, London'

valued, even then, at £197,500. One of the most lovely was a group of jewelled singing birds in a cage supported by four little gold rhinoceroses at the corners and a silver elephant in front, on a lapis lazuli stand. His famous 'Perpetual Motion' clock, over seven feet high, occupied the place of honour in the centre of the exhibition [figure 79]. It was kept going by changes in the pressure of the atmosphere. It is still in existence and came up for auction last year at Sotheby's where it was sold for six hundred pounds (one thousand seven hundred and fourteen dollars). But at the time of its invention this extraordinarily brilliant device was not taken up, and it had to wait until the middle of the twentieth century to be properly appreciated and put on the market as the 'Atmos clock'. He made another 'Perpetual Motion' clock that was actually kept going by the opening and shutting of the door of the room in which it stood.

After Cox's exhibition had been open to the public for two seasons, he obtained an Act of Parliament in 1773, authorising him to dispose of his collection by means of a lottery. The awarding of the various prizes took place in June, two years later.

In the midst of this attention to exports by certain English clockmakers, there was a distinguished import into the British Isles—a Swiss clockmaker, Justin Vulliamy, who settled in London in the seventeen-thirties where he married his English partner's daughter, and founded a horological family which continued for two centuries. He became clockmaker to the King, a position his family held until 1854. His son Benjamin succeeded to the business while Benjamin's son, who had the same name, was later to become esteemed for precision clocks [figure 80]; another descendant of this family was concerned with the original plans for Big Ben.

Meanwhile amongst other English clockmakers who concentrated on the export market were Morris Tobias [figure 81] and George and Edward Prior [figure 85], whose clocks went mainly to Turkey [figure 71], and it is incongruous to our western minds that the Sultan should have ordered a musical clock which played, at his special request, *The Bluebells of Scotland*.

82 A water-clock made at Cheapside, London, in 1735

Wherever anything lives, there is, open somewhere, a register in which time is being inscribed.
HENRI BERGSON

Domestic Clocks 1700-1961

83 (*above*) A seventeenth-century night clock from France

84 (*above right*) An American mahogany mantel clock of 1835 made by Birge and Fuller of Bristol, Connecticut, height 26½ ins. It has an eight-day waggon-spring movement, a white painted dial and painted glazed panels in its Gothic double steeple case

Remember that Time is Money
BENJAMIN FRANKLIN

BEFORE THE INVENTION OF matches some clocks were especially made to show the time at night [figure 86]. Samuel Pepys, in his diary of 1664, wrote: 'After dinner to White Hall; and there met with Mr Pierce and he showed me the Queene's bedchamber, with her clock by her beside, wherein a lamp burns that tells her the time of the night at any time'. In the seventeenth-century, the most typical night clock had a top which lifted off so that a lamp could be put inside [figure 83]. The light would shine through a small hole which lit up a revolving disc showing the time of night. A more spectacular effect was obtained with a clock which projected a magnified reflection of its face onto the wall, in the manner of a magic lantern [figure 91]. Night clocks without lights included repeaters that chimed to the nearest hour at

85 A tortoiseshell and ormolu clock of the last quarter of the eighteenth century, signed 'George Prior, London'. Above the face there is a small marine view

86 This late seventeenth-century night clock once contained a candle to illuminate the dial

87 An astronomical clock made by W. Wright of Aberdeen for the Great Exhibition, 1851

the pull of a cord; and a fifteenth-century type with a horizontal dial on which the numbers were raised, so that by touching them you could feel the time.

American clocks made before 1750 are mostly in museums and rare, such as the roughly hewn long-case clock of about 1720, which is working to-day, now at Claverton Manor, in the American Museum in Britain. The most revered single maker is David Rittenhouse, who practised from 1750 to 1790 in Morristown and Philadelphia. A large grandfather clock of his, with a case in the Chinese style which owes much to Chippendale, is one of the finest antique clocks in America. The two most outstanding families of American clockmakers are the Willards and Terrys. Simon Willard, who patented the popular banjo shaped clock, was the most celebrated of four clockmaking brothers who worked in and around Boston in the seventeen-eighties. Eli Terry of Connecticut in 1803 was making wooden-wheel clocks with movements not unlike those made in the Black Forest in Germany. He it was who brilliantly exploited the French watchmaker Frederic Japy's idea of making all the parts of a clock interchangeable, so that instead of finishing one clock before starting the next, any number were put together from large quantities of identical parts on an assembly line. Thus Eli Terry was the first maker anywhere in the world to mass-produce a complete clock. The early wooden movements were sold for around fifteen dollars. These were later superseded by brass ones, which were just as popular and as cheap to produce. Connecticut shelf and wall clocks [figure 94] were sometimes powered by small wagon-springs, which were connected by gut to the fusee. Invented by Joseph Ives, they were mostly made by Birge and Fuller from 1825 to 1855 and they are now hard to come by [figure 84]. Soon it became possible to export large numbers of these strong and cheap thirty-hour American clocks to Europe, where they flooded the market and upset the trade, practically putting the Scottish clock-making trade out of action altogether. Many of these cheap imported clocks can be found today in European antique shops and markets. They may have texts and mottoes on them, such as 'Hark! What's the cry', 'Prepare to meet thy God today'. Concurrently for the American home market a large number of quality clocks were produced with clock cases not only in the European tradition but also in a splendid variety of adventurous new designs such

88 An early nineteenth-century French Empire mantel-clock surmounted by a figure of Ulysses

as the decorative shape of acorns [figure 89], or set in the body of a realistically painted minstrel figure [figure 96].

In Paris meanwhile, about 1823 Raingo was producing a clock, on the top of which was a contrivance actuated by the clockwork, to show the movement of Mercury, Venus, Earth, Mars, Jupiter and Saturn, for which he

89 An American 'Acorn' eight-day mantel clock, built by the Forestville Manufacturing Company, Connecticut, in 1850; height 24¼ ins

90 A nineteenth-century gas clock by Pasquale Andervalt of Trieste, now in the Clockmakers' Company Museum. Its motive power was hydrogen

91 An English mantel projection night clock of 1810, signed 'Schmalcalder'. This has a twenty-four hour movement and at night projects the time backwards onto a wall. The case is of red japanned metal

got many requests. Thus several have survived in the Soane Museum, the Musée des Arts et Métiers, the Glasgow Museum, and at Windsor Castle. In England this type of clock is called an Orrery under the chauvinistic and mistaken impression that the Earl of Orrery was the first man to commission it.

In previous centuries domestic clocks had been com-

92 A French mantel-clock, *c.* 1880, of marble with ormolu mounts

parative luxuries [figure 97], but by the nineteenth century they had become commercially developed, cheap enough to be within the reach of everyone. Thus advertisements in newspapers were not unusual: 'Job Rider, Belfast, has commenced business at The Reflecting Telescope in Shambles Street where he makes clocks of all kinds to the common manner'. This was particularly so with mantel

93 *The Clockmaker,* a mid-nineteenth-century engraving by G. Hoesch

94 An improved American eight-day shelf clock with brass movement, made and sold by E. N. Welch of Forestville, Connecticut. It has a view of the new Opera House at New Orleans painted on the case

95 (*above and near right*) A free-pendulum clock and 'slave' by W. H. Shortt

clocks which were made in any number of fascinating shapes and sizes; sometimes like small cabinets, other times like little marble monuments, or with famous historical or mythological figures [figures 88 and 92]. One can see the sort of thing painted in Degas' *The Bellelli Family* or in Manet's picture of his wife at the piano, which are both in the Louvre. There was a tremendous fashion for

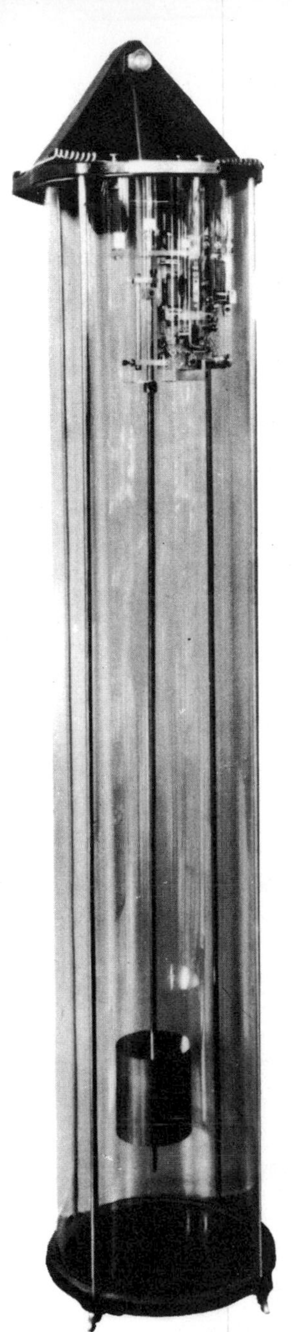

96 An American Nigger Minstrel 'Sambo' clock of 1875, height 15½ ins. It has a thirty-four hour movement and its white paper dial is set in a cast-iron painted case with moving eyes

98 A caricature by E. M. Woodward, *c.* 1820. The stupid footman, told by his master to bring him the time digs up a sundial

97 A Swiss gold and enamel box containing a clock and a musical box, *c.* 1770

clocks under glass domes from the 1830's onwards, some of which showed the movements of the mechanism, while others rested in little landscapes, in front of which a tight-rope walker might perform or a boat might sail. An extension of this idea were paintings of churches and towers, in which clocks were let into the picture from behind the canvas, instead of the painted representation. To catch the public fancy, celebrated artists at this time were commissioned to design either attractive or amusing clock-faces [figure 100]; and some of them did popular caricatures [figure 98] as well as sentimental paintings [figure 93] about Time, and about timekeepers and their makers and menders.

While astronomical clocks continued to be made [figure 87], another kind of clock with several dials, showing the time simultaneously in different parts of the world, was a successful innovation. Kings and Presidents would have them, and were thus able to see the time in all the principal cities under their rule. Similar clocks were made for public interest also. In Bernberg, there is one telling the time in Algiers, Athens, New York, Copenhagen, Istanbul, Frankfurt, Munich, Madrid, London, Berlin, Paris, Vienna, Rome, Calcutta, Peking, Philadelphia, Jerusalem, Mexico, Leningrad, and no doubt all points west of Kamskatcha too. Finally, there is the kind of timepiece which shows a map of the world across which passes a moving time-band, with

Roman numerals, giving the hour everywhere. On the other hand an unsuccessful idea, inspired by contemporary chemistry, was the clock made by Pasquale Andervalt of Trieste which was worked by gas and could be dangerous. Zinc pellets which dropped at regular intervals into hydrochloric acid produced hydrogen gas which motivated the clockwork. One of these fantastic objects, painted scarlet, is now featured at the Clockmaker's Museum at the London Guildhall [figure 90]. The manufacture of clocks continued, as before, to be divided into three categories, utility, luxury, and precision. Probably the most luxurious were the richly and finely designed clocks created by Carl Fabergé, chiefly for the Czars and the Imperial Russian Court. To Fabergé's mind, it was not enough that his clocks should be covered with the finest of precious stones, gold and enamel [figure 101]. He insisted that they should also contain the finest craftsmanship available, and thus they achieved the status of minor works of art. Sharp contrast in design is shown between these profusely and richly patterned clock cases and to-day's aggressively plain utility clocks which look, if designed by the Italians Mangiarotti and Marassutti, as smooth as pebbles, and have a lot in common in their shape with a ship's binnacle [figure 99]. From the precision point of view, between 1921 and 1924, W. H. Shortt made an electric clock in which the pendulum, absolutely free of the slave clock, became a more accurate timekeeper than ever before [figure 95]. This added the inventor's name to the list of men whose ideas have had an important effect on clockmaking. But electricity in its turn was superseded by a new and better way for achieving time accuracy. This was to measure the vibrations of crystals of quartz. Briefly this means controlling the frequency of an electric current by means of the harmonic vibrations of crystal in the same way that a crystal was used to control frequencies in radio. By 1942 quartz crystal clocks had replaced pendulum ones in great observatories. Whereas the pendulums erred a few thousandths of a second every day, the quartz crystals erred ten times less than that. But in recent years these clocks have been surpassed by a timekeeper based on atomic vibrations. A clock error of only one second in three thousand years has been achieved. Thus scientists are already considering the possibility of defining time in terms of the atom and not any more in terms of the movement of heavenly bodies.

99 An Italian table-clock designed in Italy by Mangiarotti and Marassutti in 1961

100 A print of an early nineteenth-century French clock face; *The Gossips*, attributed to Gavarni, 1825

101 This Chanticleer Egg clock by Carl Fabergé was given to the Dowager Empress Marie Feodorovna by the Czar Nicolas II in 1903. Its case is made of translucent blue enamel on a shimmering *guilloche* surface of gold. At each hour the cock appears, crows and flaps its gold and diamond wings

102 Thomas Mudge's celebrated Number 1 chronometer, width 5 ins, was a contestant for the Board of Longitude's award of £10,000

Time and Navigation

Time is as wind, and as waves are we.
SWINBURNE

A SHIP'S EXACT POSITION at sea can be estimated so easily to-day that it is difficult to realize that only two hundred years ago it was impossible to do it with any real accuracy at all. Because of the growth of import and export markets, and the great voyages of exploration in the eighteenth century, it became more and more important to keep ships directly on their course. The nation which could invent a device to do this would immediately have an immense advantage over its commercial rivals. And it all depended on clocks [figure 109].

The Dutch scientist Gemma Frisius, as early as 1530, had held that if one could take a reliable clock on board ship, set to the time at home, say Greenwich or any other starting point, and when one were far out at sea one compared its midday time with one's ship's noon time (known from the sun's position), then the difference between these times (if one calculated that one hour's difference was equal to fifteen degrees of longitude either east or west of home) would give one the precise position. But in the sixteenth century portable clocks stopped ticking at sea and in any case were not reliable enough to enable precise calculations to be made. Then, sailors could find out the exact time of day at sea from the sun's altitude by using an astrolabe [figure 103], but because a sea-going clock had not yet been invented to give Greenwich time they were quite unable to estimate the ship's position in the way Gemma Frisius proposed. So a clock was required that would keep going merrily at sea through storm and battle. In 1598 King Philip of Spain had offered a reward of a hundred thousand crowns for this invention. Not long afterwards the Government of Holland offered ten thousand florins. The Kings of France encouraged clockmakers in every way to evolve a seaworthy timepiece. The Paris Academy joined in with a prize. Then in 1659 at the Hague Christian Huygens put together a clock specially constructed for sailing. He used a pendulum clock attached to a contrivance for keeping it as upright as possible; it went well in calm weather, but in raging seas it stopped. One of these clocks was taken on a voyage along the West African coast without success. This voyage, however, brought one useful piece of knowledge, that the rate of the pendulum changed the nearer it got to the equator. And Huygens was baffled by the irregularities caused by changes of temperature; a hazard brilliantly overcome, as we shall see later, by John Harrison, with his gridiron pendulum made of two metals which expand and contract in opposition, and by George Graham later with his mercurial pendulum [figure 104]. The difficulties of making a reliable sea-going clock seemed insuperable. Morin, the astronomer and clockmaker,

103 An astrolabe, simplified for sailors, which was recovered from the wreck of an Armada galleon in 1588

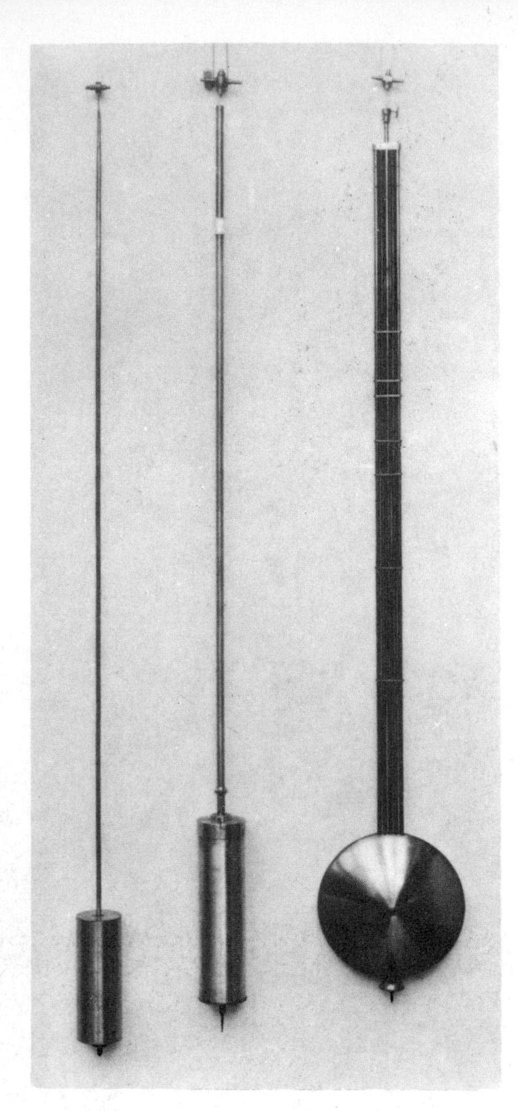

104 Three clock-pendulums, showing (*left*) the uncompensated mechanism, (*centre*) with Graham's mercurial compensation, and (*right*) with Harrison's gridiron compensation

said to Cardinal Richelieu: 'I know not what such an undertaking would be to the devil himself, but to man it would undoubtedly be the height of folly.'

The next move was in 1714, when the British Government offered ten thousand pounds for a clock that would be accurate to one degree after a voyage to the West Indies and back; with an offer of greater rewards up to twenty thousand pounds for yet greater accuracy.

A man who dedicated his whole life to chronometers, John Harrison, was born on the Nostell estate near Pontefract in Yorkshire in 1693 [figure 106]. By 1728, when he was thirty-five, he had completed his plans for a spring-driven clock that would, he believed, keep going through all weather and hazards at sea, and set off for London hoping to win the reward. His confidence was shaken, however, when he showed his designs to 'honest George Graham', the inspired clockmaker, who advised him to go back to Yorkshire and make the chronometer to see if it really worked. Graham was so impressed by what he had seen that he lent Harrison two hundred pounds. Over the next seven years, while carrying on a business in Yorkshire as a clock-mender, John Harrison experimented with various models of his design until he had perfected it. Then he returned to the capital. His Number I chronometer was put aboard a ship going to Lisbon and back [figure 105]. It was more successful than anyone expected, and though it was not perfect he was immediately awarded five hundred pounds and encouraged to work on it. This he did, but his new ideas could not be tried out at sea because England was then at war with Spain and the Government feared that the chronometer might fall into enemy hands. By the time the war ended Harrison had completed his famous Number 4 chronometer, which in all essentials was a large watch almost five inches across [figure 107]. It was put aboard *H. M. S. Deptford* in 1761 on a voyage to Jamaica and back. This was thirty years or more since John Harrison had first come to London. On this famous voyage, William Harrison the son looked after his father's chronometer. It went so well that the incredulous Board of Longitude did not trust the good result and deciding that it must have been mere chance, refused to pay John Harrison anything. Three years later another test was made, during which his chronometer gave an even better performance. The Board of Longitude now reluctantly

105 John Harrison's Number 1 chronometer, designed in 1735

106 Portrait of John Harrison, the English chronometer-maker, by an unknown artist

paid him five thousand pounds on account. After another successful trip, a few more thousands were granted to him, but it took a petition to George III in 1773 to obtain him his just reward of twenty thousand pounds. Harrison's chronometers were handed over to the nation. He died in 1776 aged eighty-three, and his tomb is in the south-west corner of Hampstead churchyard; his chronometers are in the Naval Museum at Greenwich. He had been one of the first clockmakers to overcome the irregularities in timekeeping at sea caused by the expansion and contraction of the metal parts of a clock due to changes in temperature. To-day, however, of all the methods once used to compensate for these varying temperatures, that of the French clockmaker, Pierre Le Roy, is the one now universally adopted.

107 John Harrison's Number 4 chronometer, which eventually won him the Board of Longitude's award

The Board of Longitude had amongst its members Isaac Newton, Thomas Mudge, and Lareum Kendall, who at the request of the Board made a duplicate of Harrison's invention. He then adapted this into a much simpler instrument than Harrison's, dated 1771. It is said to have belonged to Captain Cook and was taken by him to the Pacific in 1776.

The French Academy's offer of a handsome prize for a chronometer still held good. Pierre Le Roy submitted his design together with a chronometer, and so did several other well known makers, including the equally brilliant Ferdinand Berthoud, Swiss by birth and French by adoption. Le Roy's design was looked on favourably, and the Academy somehow managed to lose Berthoud's plans. Le Roy's chronometer was never tested at sea, so nothing came of it, and the prize was withdrawn. Now for a time both men went on experimenting. In 1768, two of Le Roy's sea-clocks were tested in a voyage of a hundred and sixty-one days to Newfoundland, with fair results. Three years later, two chronometers, one by Le Roy, the other by Berthoud, were tested on a year's voyage to Iceland, Newfoundland, and Copenhagen. Le Roy's, which was doing very well, was damaged after fifty-four days. Berthoud's went more or less satisfactorily. It is clear there was little to choose between the two, but what difference there was went in Le Roy's favour, and he therefore had every reason to be upset when his work was completely ignored in a book claiming to give a complete account of work done in France on chronometers, written by Ferdinand Berthoud and published in 1773. It led to a famous quarrel between the two brilliant men. These enemies glare at each other to-day in a corridor of Buckingham Palace, where a tall clock by one faces a tall clock by the other.

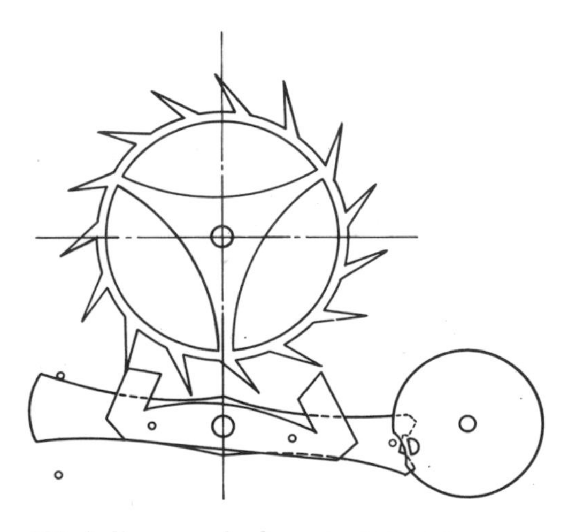

108 A diagram of a lever-escapement

Meanwhile the French Academy had doubled its former offer and the British Government announced fresh rewards for a more easily produced chronometer than Harrison's. A man who applied for the latter was that jovial west countryman, the famous clockmaker Thomas Mudge, who had been one of the members of the Board of Longitude that had examined and approved John Harrison's prize-

109 *The Measurement of Time and Navigation* by G. N. Nasini [1657-1736]

110 Wheatstone's helio-chronometer, now in the Science Museum, London

winning instrument. He had already invented the lever escapement [figure 108] and successfully applied it to a watch for Queen Charlotte. This escapement, so-named for the very good reason that its escape mechanism worked with a lever, caused watches to be absolutely immune to irregularity from constant changes of position and the joltings of coach travel. And this lever escapement is still to-day generally used in watches. Mudge's own chronometer [figure 102] had proved its worth on a trip to Newfoundland and back, but when he applied for the promised reward he, like Harrison before him, had the greatest difficulty in obtaining it. His Number 1 chronometer was kept at Greenwich for fourteen years, while being tested for accuracy. It was not until he had petitioned Parliament and brought the matter before a select committee of the House of Commons that he was eventually awarded his three thousand pounds. In 1743 his chronometer was sold at Christie's for eleven guineas by Queen Victoria's impecunious uncle, the Duke of Sussex. It is now to be seen with the Ilbert Collection in the British Museum. The Government's slowness in paying Mudge was due to the fact that the Astronomer Royal, Dr Maskelyne, had found a chronometer he thought better than Mudge's, which had been sent to Greenwich also but not yet tested. It had been made by young John Arnold, who had a clock-making business in the Strand. Captain Cook, who was one of his customers, took Arnold's Number 3 chronometer with him on his voyage of 1722. But rightly or wrongly, it was said that Arnold's fine chronometer had been stolen from designs made by a rival young clockmaker, Thomas Earnshaw, who in 1803 was awarded three thousand pounds for his chronometer. It was the improvements made to chronometers by these two men that made them commercial propositions. Great improvements had been made in instruments, making it easier to tell the time of day at sea from the sun's altitude. Outstanding amongst these was Wheatstone's helio-chronometer [figure 110]. The time was indicated on a dial with hour and minute hands (as on a watch), fixed to the lower end of a brass tube with a lens at the top. When this tube was so adjusted that the sun shone through the lens, then the hands of the dial showed the correct time. It has been well said that it was natural that Britain, being a sea power, should devote more attention than other nations to chronometers.

Japanese Clocks

111 A Japanese striking clock in an etched brass case

NOWADAYS IN JAPAN CLOCKS are much the same as they are elsewhere, but before 1873 (when the Japanese adopted western methods of time keeping) their clocks were entirely different, a branch of horology on its own. Up to that date Japan had evolved her own native way of time reckoning [figure 115]. The complete day was divided into two parts, from sunrise to sunset, from dusk to dawn. Each part was given six 'hours'. This meant six hours to the daylight section, and six hours to the night. But as both day and night by their very nature vary in length with the seasons, it follows that an hour in one part could seldom be of the same duration as an hour in the other. Thus in high summer a daylight hour could be as much as two-and-a-third times as long as a night hour, (and in that event one Japanese hour was equivalent to two western hours).

An early mechanical clock to reach Japan was a Spanish one, of the brass lantern type, made by Hans de Evalo, which was taken there either by Portuguese or Dutch traders. Japanese craftsmen then grafted their own way of time reckoning onto European mechanical clocks, displaying a remarkable flair for improvisation. The first way Japanese clockmakers adapted western clocks to portray the irregular Japanese hours was by a simple mechanical adjustment to be made at dusk and at sunrise. This was a tedious business. A far better idea was to have the hour numerals on the dial marked on little moveable plates. It was enough, if these plates were altered once a fortnight by a clockman, for them to indicate the continually expanding or contracting hours with some accuracy. In midsummer the dial would show the hours unevenly spaced; with a crowded group of six for the short summer nights; and a widely spaced group of six for the long day hours. There were others with an altogether different *modus operandi*, which had two adjustable regulators, one for night, one for day, which automatically switched from one to the other at sunrise and at sunset [figure 111].

The actual hours in Japan were also counted in a different way. Not only was this done backwards, but the

112 A Japanese print, showing a table clock

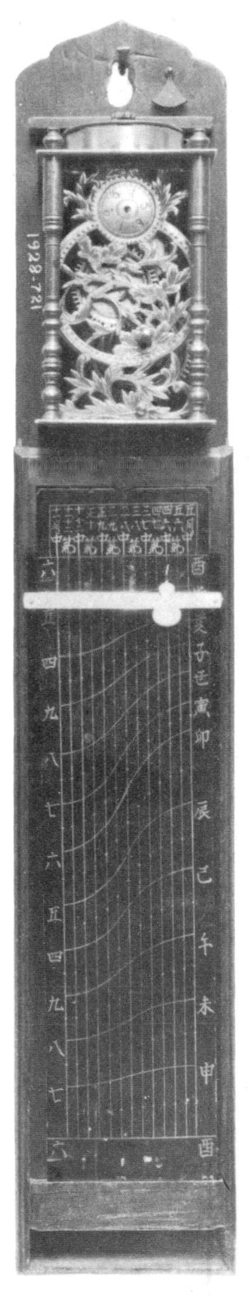

113 A Japanese pillar clock controlled by balance and spring

numbers 1, 2, and 3 could not be used for clocks as they were sacred, reserved for religious use only. So the numbers 4 to 9 had to mark the six hours in each part of the day; each dial therefore showed 9, 8, 7, 6, 5, 4, repeated, with 9 at the midday and midnight positions. To tell night hours from day hours (instead of using AM or PM), each numeral was given an animal equivalent: nine at midday

114 A Japanese print showing a wall clock with dangling weights

115 A Japanese print showing a pyramid clock

a Horse, at midnight a Rat; eight in the morning an Ox, in the evening a Sheep; seven in the morning a Tiger, in the evening a Monkey; six in the morning a Hare, in the evening a Cock; five in the morning a Dragon, in the evening a Dog; four in the morning a Snake, in the evening a Boar.

These weight-driven Japanese clocks are of four different types. Three of them, somewhat similar in size and structure to European lantern clocks, were made from the early seventeenth century onwards. The first, hung by a silk cord or hook, was the wall-clock [figure 114]; the second, mounted on a lacquered truncated cone, down which the weights could hang unseen, was the pyramid type [figure 115]; while the third, placed on a table, often with cabriole legs, had a hole in the top for the weights to go through, and was known as the table clock [figure 112]. A fourth variety of weight-driven clock was also invented. Designed to hang unobtrusively on the central pillar of the house in a sparsely decorated Japanese room, these pillar clocks were from one to four feet tall. They were the cheapest of Japanese clocks, but unique because, unlike any other clock, they told the time vertically [figure 113]. A hand fixed to the weight as it descended showed the time on a rod marked with adjustable hours. There are many variations on this vertical theme, all of a wonderfully effective simplicity. Most Japanese clocks strike the hour, and in some cases strike the half hour as well. It is done in this way: two strokes mark the half hour after an even-number hour, but only one stroke after an odd-number hour.

It was in this same year, 1839, according to the American collector James Arthur, that spring-driven clocks at long last made their appearance. Some of them had a fixed hand which pointed straight upward while the dial moved round; others had the more usual arrangement. In shape, eight or nine inches high, these spring driven timekeepers looked not unlike a Tompion bracket clock; yet had glass in the sides as well as the front; the woodwork vas glued together. One of their most famous makers was Denjiro Kobayshi of Tokyo. The first factory for modern clocks in Japan opened in 1875. After that date old-fashioned models were either thrown away or exported as antiques to Europe and America, where they have become collectors' pieces. It is more or less impossible to find any of them now outside museums in the Orient.

Monumental Clocks

Be ruled by Time, the wisest counsellor of all.
PERICLES

THE FIRST MONASTIC clocks were made to strike an alarm rather than show the time on their dials; probably many had neither dials nor hands but told the time only by bells. They resembled the mechanism of the Salisbury Cathedral Clock (1386) now standing on the floor of the north nave [figure 119], which may well have been made by the same hand as the one high up in Wells Cathedral, Somerset [figure 118]. The latter is of the same date, but has a dial. So had the first public clock about which we have any definite information; it was put up in 1335 at Milan. This had a twenty-four hour dial and would strike up to that number, ending after sunset. Soon public clocks were being built all over Europe; in Seville, Moscow, Padua, Bourges and in England at Wimborne in Dorset, at Ottery St Mary, and at Exeter. These clocks, often built in churches, were highly treasured. Froissart the historian says that Philippe the Hardy, Duke of Burgundy, after victoriously entering Coutrai, had its town clock removed and transferred to his own duchy at Dijon. In 1335 King Edward III of England imported some Dutch clockmakers to make a clock for the tower he had built at Westminster, near where Big Ben now stands. The

117 The chimes played by Big Ben

116 Big Ben, London, a photograph taken while workmen were cleaning one of the four faces

118 An engraving of the Wells Cathedral clock, originally built for Glastonbury Abbey in 1392 and moved to Wells under Henry VIII. The dial is approximately six feet across. This mechanism was repaired in 1960 by London's oldest turret-clock-making firm, Thwaites and Reed, founded in 1740

result was that the citizens complained that the bells soured the beer in the local taverns, in rather the same way that Londoners believed that Big Ben attracted zeppelins in 1915. Nothing of this Westminster clock remains; nor of the first monumental clock in France, which was on the tower of a fortified bridge at Caen; but a few parts of the thirty-six feet high clock in Beauvais Cathedral, which had fifty dials, are still in evidence.

119 The mechanism of the Salisbury Cathedral clock, built in 1386

South German makers excelled at monumental automaton and astronomical clocks. At first the automaton was a single Jack figure who struck the bell, symbolizing the town crier who used to go round calling out the hours and who had been ousted by the clock [figure 121]. One of the best examples of these Jack strikers is on the famous *Zytglogge* at Berne. Soon short scenes were being enacted. In 1442 a clock was built in Nuremberg with

120 The clock on the tower of the Palais de Justice in Paris

figures of soldiers marching in and out hourly; known both then and now as the *Maennleinlaufen*, the running of little men [figure 122]. At Hamlin robot children followed a Pied Piper; while at Lund in Sweden in the early days two Knights appeared when the clock struck, and struck each other (instead of a bell) as many times as the hours decreed. But to us nowadays, who can see the whole world in a flash on the television screen, one automatic clock is very like another. In fact when you have seen one you have seen them all. Yet when these automatic clocks were originally put up on public buildings, each was a sensation. They became one of the sights, and excited passers-by stood and stared. Provincial visitors, for instance, to London would stop to watch the two giants with clubs in their right hands strike the bells of St. Dunstan's-in-the-West in Fleet Street.

It would seem that the passion for astronomical clocks long ago came about because man believed that by exactly copying the routine of the heavens in this way, he might possibly discover their secret. How else can one explain the fact that nothing would stop mathematicians, astronomers and clockmakers across Europe competing in designing new mechanism more and more elaborately? Yet these clocks were still far from accurate. The Clergy delighted in them, not only because of the way in which they brought the heavens to the church, but also because they had calendars marking Saint's Days and Feast Days. To the congregations these astronomical clocks seemed to confirm their beliefs about the universe, and, together with stained glass and sculpture, painting and incense, seemed to be another of God's wonders duly performed.

121 An English sixteenth-century figure of a Jack, which was used to strike the hours

122 The *Maennleinlaufen*, a large musical clock in the Church of Our Lady at Nuremberg. The work shows the Emperor Charles IV with seven Electors doing him homage

The great prestige that went with these monumental clocks appealed also to municipal pride. Town Halls and ministries boasted them. At the invitation of Charles V of France in 1370, the German scientist Henry de Vic went to Paris to supervise the erection of a great clock on the Palais de Justice [figure 120]. When it was first set going, Parisians compared temperamental people to it, saying; 'He is like the Palace clock and only works when he wants'. Another great medieval artifact, still working, stands at Rouen [figure 127]. Built in 1389 it is the earliest clock known to strike the quarter hours, while its dial was the first to be marked with hours from one to twelve, instead of the older way from one to twenty-four. In 1483 such a wonderful astronomical and automaton clock was made to adorn the Town hall at Prague, that scientists wrote lengthy eulogies about it. This fabled clock, destroyed in the last war, has now been rebuilt, with the essential difference that, whereas before the figures were charming bourgeoisie, they are now smiling Soviet workers. Probably the most spectacular and intricate automaton and astronomical clock in the world is in Strasbourg Cathedral. Evolved in 1570, to replace a lesser fourteenth century predecessor, to a plan by Isaac

123 A print of the second Strasbourg Cathedral astronomical and automaton clock

124 An indoor monumental clock, from a fifteenth-century Flemish illuminated manuscript in the British Museum

125 An interesting early model of the famous clock tower in St Mark's Square, Venice, which was erected early in the seventeenth century and was the second on this site. It has a rotating twenty-four hour dial, decorated with the signs of the Zodiac, height 10 ft 5 ins

126 The church clock at Rye was presented by Queen Elizabeth I to the town. It was later given a pendulum which hangs right down into the nave to within twenty feet of the floor

127 An old print of the tower-clock at Rouen, built in 1389 and still working

Halbrecht, the splendour of this work of art can be seen in the illustration [figure 123]. 'Celebrated next to that of Strasbourg,' wrote the English diarist and traveller, John Evelyn, in 1645, 'for its many movements, is the tower clock in St Mark's Square, Venice.' A splendid model of this clock is illustrated in these pages [figure 125]. Some alteration to its workings were made in 1859, so that numbers showing the hours and minutes appeared on either side of the Virgin instead of the Saints. Every summer crowds of holiday makers stand watching its manipulations while the Jack figures on top strike the

128 The church clock erected by Manoel Pinto de Fonseta [1741-73] in the Grand Master's palace at Malta. The bells are surmounted by two figures of Malta's traditional enemies, the Moors

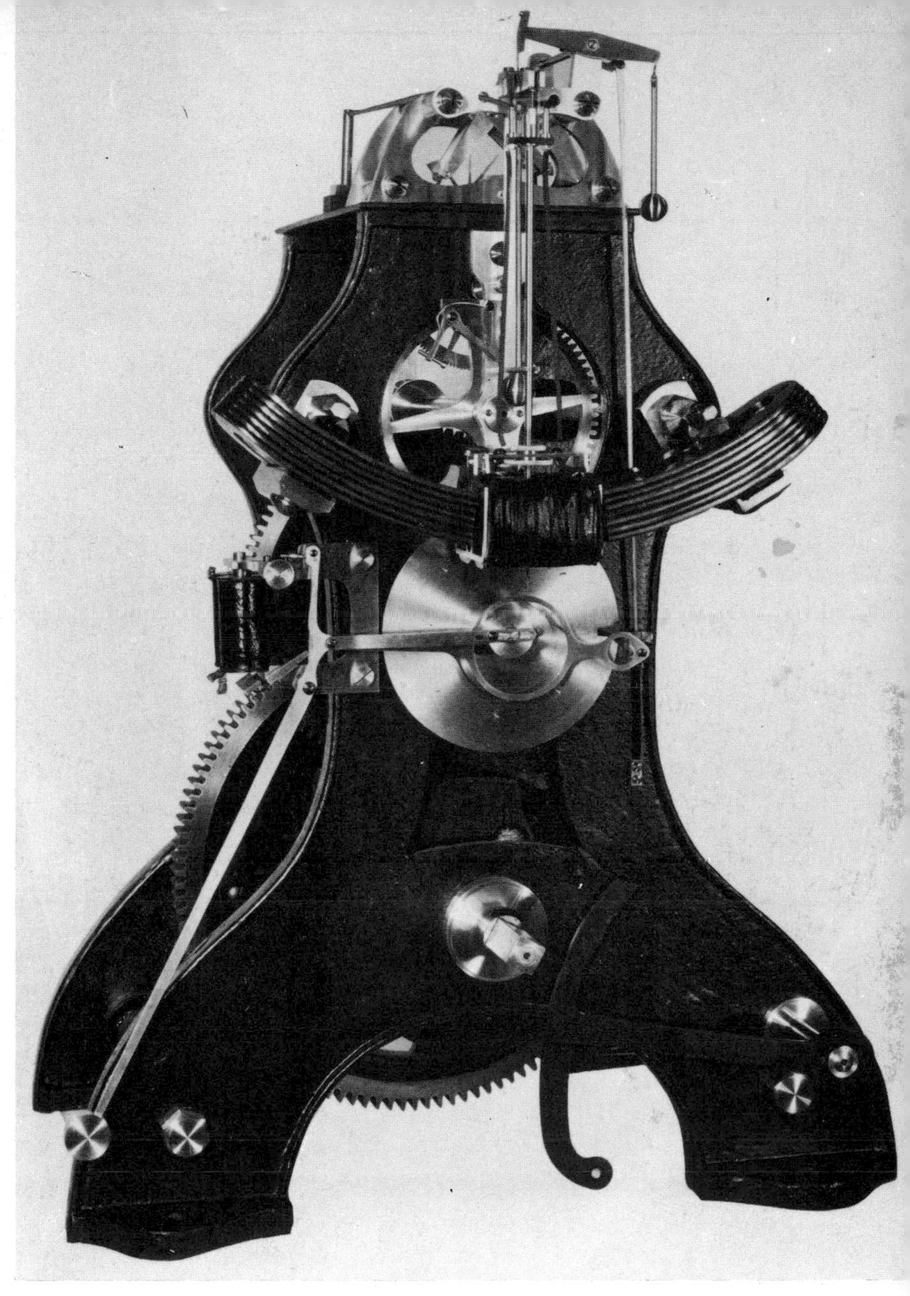

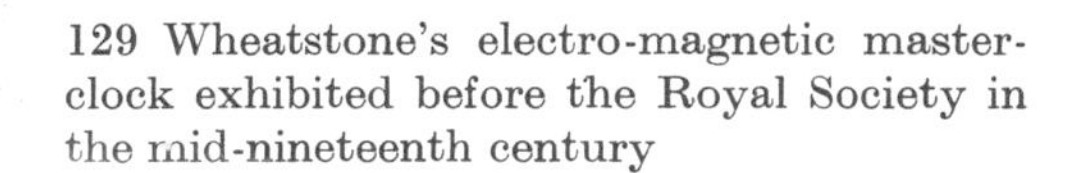

129 Wheatstone's electro-magnetic master-clock exhibited before the Royal Society in the mid-nineteenth century

bells and chime their holidays away. The great variety of public clocks erected in the mid-eighteenth century can be seen from the architecturally splendid automaton clock put up on the island of Malta by the much-hated ruler Pinto [figure 128]; and from the indoor monumental organ clock, with a case by the great cabinet-maker Roentigen, which marks the passing hours by automatically playing pretty tunes [figure 133].

The first tower clocks in America were neither astronomical nor automaton. Freed from these traditional embellishments, their large black-painted faces with golden

130 An astronomical clock on the Financial Times Building in London, which was set going in 1959. Its builders, Thwaites and Reed, expect that 'it will be quite all right without regulating for a hundred and fifty years'

131 The modern clock on the Town Hall of Wolfsburg, Germany, with its bronze carillon bells. The chimes are among the most beautiful in Europe

Roman numerals told the time boldly to the New World. The Boston public clock of 1657 has now disappeared, and Samnal Adams Drake in *Old Landmarks of Boston* considers that these 'clocks' were in fact bells. In May, 1716, at a public meeting in Boston there was a discussion about providing a town clock in the Brick Meeting House for the benefit of the inhabitants. An eight-day clock was therefore erected by Mr Benjamin Bagnall in 1717, according to the *Boston Selectmen*, but there is some doubt as ot whether this date is correct. The turret clock at Guilford, Connecticut, erected in 1726 [figure 132], continues to work and is the oldest monumental clock in action in the United States to day; however, it is indoors instead of out. Subsequently many tower clocks were made by Simon Willard, the maker famed for his 'Banjo-clock' inventions. A clock crowned with the statue of William Penn was erected on the Philadelphia Town Hall in 1873.

But the largest monumental clock of the nineteenth century was built in 1854 for the Houses of Parliament, London, with its famous bell, Big Ben [figure 116 and 117]. It too is neither astronomical nor automaton. In 1844, Vulliamy, of the third generation in a great family of clockmakers, was commissioned to design a great clock in the new Houses of Parliament. A rival, E. J. Dent, who had just put up a tower-clock at the Royal Exchange, asked to be allowed to compete for the making of the clock, and eventually his plans were chosen. Dent, however, had died in 1855 before the clock was built, so the supervising of its making was left entirely to the designer of the first really successful tower-clock escapement, Lord Grimthorpe. Vulliamy had many influential friends; their opposition and the incidental difficulties during the building would have caused any less determined man than Grimthorpe to abandon the project. The hour bell cracked in August, 1856. A second bell, which replaced it in 1858, soon acquired a fissure where the hammer hit it. But oddly enough when this same fractured bell was twisted round slightly, so as to present a different surface to the hammer, it worked splendidly; and has done so ever since, and was given the name of Big Ben in honour of Sir Benjamin Hall, who commissioned it for the Ministry of Works.

At the two great European exhibitions of the nineteenth century, one in London (1851), the other in Paris (1867), amongst the array of horological beauties and beasts were some commercial tower clocks, in which one single mecha-

132 The oldest tower clock in America, still working at Whitfield House, Guilford, Connecticut

nism propelled the hands on a large round clock face intended for the outside of a building, and the hands of a suitably smaller face for the inside of it. Exhibited also were clocks mounted on lamp-posts from the Breguet firm, worked by compressed air and known as *horloges pneus*. In Spain other lamps of the same type decorate the Promenade at Jerez de la Frontière. They are marked J. R. Losada, Regent Street, London. Losada was a Spaniard who came to London in 1835 and started up as a clockmaker, doing a large business chiefly with Spain and South America. Electric clocks were exhibited for the first time at the London and Paris exhibitions. In 1840, Alexander Bain, a Scotsman, patented some ideas for electric clocks, the main one being that one particular clock should send out electric currents every second, minute and hour, thus controlling almost any number of distant clocks, which would all keep exactly the same time together. This system was not an immediate commercial success. Hipp of Neuchâtel experimented with this idea, producing a magnetic impulse clock system. As the result of adaptation of the whole idea by various other inventors [figure 129], synchronising methods like it are now used in factories, municipal buildings, and over whole railway and aircraft systems.

133 The clockwork is by Christian Mollinger of Berlin [1754-1826], which tells the time on the marble globe on top of the cabinet designed by Roentigen containing an organ. It is here pictured in the Sotheby's saleroom, London, in 1961

Early in the twentieth century, as if to redress the balance between intensified mechanisation and natural forces, a new kind of water-clock, perhaps more ingenious than any of that kind evolved in ancient Egypt, was erected and successfully set in motion. Crude in construction, high in accuracy, it tells the time perfectly in the turret of the gatehouse at Kilruddery Castle in county Wicklow in Ireland. It was invented in 1898 by a young officer in the Irish Guards, who afterwards became the Earl of Meath. In it a perpetual thin stream of water falls onto a rod, fixed at right angles to the pendulum, and onto another rod lower down, fixed at an angle of forty-five degrees. The weight of the water falling on these two rods starts the pendulum swinging and keeps it going, as the rods attached to it move to and fro out of the stream and back. Only a frost or dead flies in the water supply pipe ever interfere with this clock's reliability. There has recenty been a revival in public clocks with automata mechanism. In Messina Cathedral, whose rebuilding was completed in 1930 after the great earthquake early in this century, there is a clock which is nothing less than an extravaganza of mechanical scenes and figures,

134 A unique clock-tower in Tokyo honours the 'Punctual Statesman', the late Yukio Ozaki, who was never late for an appointment in his life

built by the descendants of the apprentices of Schwilgué, who was noted for his work on the third adaptation and renewal of the Strasbourg epic clock. The fact that people stopped to stare at such clocks appealed to Big Business as a form of advertising. So automaton clocks, formerly patronised by the Church and municipalities, were now taken up by commerce. Almost every great city has several, which advertise one product or another. In London useful examples are working daily on Liberty's shop in Regent Street, and on the Houndsditch Warehouse in the City, where Gog and Magog strike the bells; while in summer in the Festival Gardens at Battersea the Guinness Clock works away merrily for all to see. This is the apotheosis of all automaton clocks, being powered by no less than fifteen electric motors.

Advertisers also came to use non-automaton clocks, by painting the name of a product across the clock face, or round the dial instead of the numbers, or by using neon signs. In America, an advertising clock was erected on the Colgate Toothpaste Factory as early as 1906 and there are others on the Metropolitan Life Insurance Building and the Paramount Building, both in Manhattan. And prestige comes into it. For why else should trouble have been taken, at the start of the century, to design the clock on the Ankar Insurance Building in Vienna in a style so impressively *art nouveau*? And what seaside resort and spa does not possess its flower-clock, carefully planted out with begonias and lobelias, the pride and joy of the municipality? At Easter in 1955 the largest bank building in Lecce, Italy, acquired a tower clock designed by the sculptor Francesco Barbieri. Around the face are twelve allegorical figures of Fortune, Strenght, Justice, Love, etc. This is certainly a prestige piece; so too is the monumental clock inside the *Time and Life Magazine* offices in Bond Street [figure 135] beautifully and wittily designed by Robin and Christopher Ironside. Father Time, the British lion and the American eagle support it. Of new municipal clocks built recently, the carillon in the tower of the Town Hall in Wolfsburg, Germany, is as good as any in the contemporary style [figure 131]; while at Ulm, (like Wolfsburg an historic clock-making centre), the new railway station boasts a public clock without numbers on the dial. Instead it shows the letters of the popular tongue twister *In Ulm, um Ulm, und um Ulm 'rum'*, in Ulm, around Ulm and round about Ulm. This was set going in 1960. Clocks

135 One of the most attractive small monumental clocks of recent years is in the *Time and Life Magazine* offices in London. Designed by Christopher and Robin Ironside, it is partly allegorical, partly heraldic, supported by Father Time, the American eagle and the British lion

like these now make the straightforward faces of public clocks with numbered dials, like the one Chirico painted in 1911 in his *Enigma of Time*, seem almost rigid, if not formal. In Tokyo a clock tower as thin as a plank and as high as the sky has been built near the Parliament, at a cost of twenty thousand pounds (sixty-four thousand dollars), to honour and symbolise the punctual character of one of Japan's greatest statesmen, Yukio Ozaki, who was never late for a single appointment [figure 134].

And strange as it may seem, fine public astronomical clocks are still being made today. Apart from the new one at Prague, the Deutsches Museum in Munich boasts a beauty. Another installed at York Minster in 1955 is the first to show the celestial motions as seen by airmen; whilst the newest in London, over the main entrance of the Financial Times Building in the City, has in the centre of its design a glowing portrait of Sir Winston Churchill represented as Phoebus the Sun-God [figure 130].